The Synergistic Effect of Sustainable Business Practices on Corporate Performance

Avichal Sharma • Majo George
Pallavi Pandey • Vivek Joshi • Manjari Sharma
Editors

The Synergistic Effect of Sustainable Business Practices on Corporate Performance

Springer

Editors
Avichal Sharma
School of Sciences
Christ University
Pune, India

Pallavi Pandey
School of Commerce, Finance and Accountancy
Christ University
Pune, India

Manjari Sharma
School of Commerce, Finance and Accountancy
Christ University
Pune, India

Majo George
Faculty of Economics
University of Economics and Finance (UEF)
Ho Chi Minh City, Vietnam

Vivek Joshi
3E- Excess Edge Experts
Jaipur, India

ISBN 978-981-96-3374-6 ISBN 978-981-96-3375-3 (eBook)
https://doi.org/10.1007/978-981-96-3375-3

This Springer imprint is published by the registered company Springer Nature Singapore Pte Ltd.
The registered company address is: 152 Beach Road, #21-01/04 Gateway East, Singapore 189721, Singapore

Foreword

I am honored to lead you into this wonderful collection of essays that discuss the economic impact of sustainable business practices. Environmental scientists were preceded by naturalists. They all have been solid advocates for sustainability for centuries. They hold that harm to the environment has no benefit for anyone. Judicious use of resources and avoidance of harm are essential for ecological balance.

The questions examined by these essays delve into the sustainable practices of business in support of economic balance. Businesses easily consider their explicit costs. Financial statements and account ledgers readily reveal those explicit costs. However, the ledgers ignore the unseen costs, the so-called implicit costs of damage to the surrounding environment. In his excellent book "Pricing the Priceless" (Cambridge University Press, 2024), H. Spencer Banzhaf recounts the growing role of economics in environmental science. The key to the sustainable balance is in the title. Pricing is always the challenge when there are no implicit costs measured on the balance sheet. What is the value of clean air to breath? What is the value of clean water to drink?

A short examination of the role of the corporation in these matters is worthwhile. For decades, the Friedman Doctrine guided business decisions. In his September 13, 1970 essay in The New York Times, Friedman boldly stated that a business exists to make a profit that thereby increases the wealth of the business owners, its stockholders. Observing and compensating for implicit costs would certainly detract from profit maximization. And thus, the corporation's purpose was clear.

Oddly, however, about the same time as Friedman's essay, the United States Environmental Protection Agency was established, opening their doors in December 1970. The first Earth Day had been celebrated just eight months earlier on April 22, 1970, but this celebration had been dismissed by the establishment as a "hippie event" that wouldn't last. Today, according to the EPA, Earth Day is celebrated worldwide by over 1 billion people who give up some volunteer time for community service.

The first significant legislation after the EPA's creation was the Clean Air Act of 1970. The standards created led to the development of the catalytic converter to make automobile exhaust less harmful. The EPA estimates that cleaner air from implementation of the Act has prevented more than 200,000 premature deaths in the United States. It took regulations and new rules to get corporations to recognize the importance of some of the implicit costs.

The Business Roundtable is an association of chief executive officers of America's leading companies. About 200 firms participate in periodic discussions about business strategy and goals. According to the Business Roundtable, "the CEOs advocate for policies that create and maintain long-term economic prosperity." Since their first statement of Principles of Corporate Governance in 1978, the Roundtable consistently supported the Friedman Doctrine that "corporations exist principally to serve their shareholders." Profit maximization is the fundamental means to attain that goal.

A major shift in the stated purpose of a corporation occurred when the Roundtable issued a new "Statement on the Purpose of a Corporation" on August 19, 2019. Many CEOs has long known that the corporation added value to many others besides the shareholders. The 2019 statement effectively replaced the work "shareholders" with "stakeholders." Acknowledging that customers, employees, suppliers, communities, and environment deserved the same fundamental commitment to value as did the shareholders, the CEOs of the Roundtable made a bold shift in the Friedman Doctrine. The short one-page Statement concludes, "Each of our Stakeholders is essential. We commit to deliver value to all of them, for the future success of our companies, our communities and our country."

One of the major challenges to policy makers and environmental scientists is agreeing on how much pollution is "okay." It is straightforward to draw a supply–demand diagram to show how much pollution could be removed by embracing implicit costs. Using the pricing mechanism to add these social costs into the private (explicit) costs is an effort to realize efficiency in production while keeping the community and environment safe and healthy. Naturally, the firm will resist those social costs. This is the ongoing battle between environmentalists and corporations with government often acting as the referee. The purpose of bringing in economists is to establish a rational framework for efficiency.

Sustainability does not mean perpetual existence. I recently reviewed the history of America's Dow Jones Industrial Average that has been in existence since 1896. The first list of the leading companies contained 12 companies. Of those only a few remain in business today, mostly after significant reorganizations, acquisitions, and mergers. Now the DJIA is comprised of 30 large companies, and none of the original companies remain on the list. General Electric Company was on the original roster, but it was dropped in 2018, being replaced by Walgreens Boots Alliance, a major pharmaceutical retailer. Walgreens Boots Alliance remained on the DJIA for only six years when it was dropped and replaced by Amazon.com, Inc.

Another benefit of sustainable practices is efficiency of a system whereby the resources are used for their maximum benefit. Economists like to call this result "Pareto efficiency." The illusive concept of Pareto efficiency where no participant can be made better off without another participant becoming worse off is a challenge. Human nature gets squarely in the way of reaching such an efficiency as participants seek more for themselves at the expense of others. If a Pareto efficient result includes both private (explicit) and social (implicit) costs, it will come closer to a sustainable system.

As we look at what might help bring about better sustainable practices, it is important to note another factor besides price. That is the aspect of ownership. It has long been observed that responsible owners take care of their property better than renters. Establishing ownership is a challenge. When I was young our family visited the famous Yellowstone National Park in Wyoming. I recall that there wasn't an admission fee to get into the park. The common resource of the park was available to all without restriction, and it became overly crowded. In 1968, Garrett Hardin published "The Tragedy of the Commons" in the journal Science. He described how overuse of common resources was inevitable. Such overuse could be managed only by establishing ownership. Since then most common resources have established temporary ownership rights by adding an admission fee. My most recent visit to Yellowstone required payment of a substantial fee of $35 for the vehicle.

As you read these essays, consider the implementation methods that might bring about the great ideas of the authors. We all have great ideas about what should and could be as we seek opportunities to make our world more sustainable. We often fall short in realizing those opportunities. We trust that you will be engaged in trying to help bring about a better level of sustainability.

Wayland Baptist University
Plainview, TX, USA

Charles N. Starnes

Preface

The twenty-first century has ushered in an era where the convergence of sustainability and economic performance has become not only a topic of interest but a crucial consideration for businesses worldwide. Historically, the notion of sustainability in business was often relegated to the domain of corporate social responsibility (CSR) initiatives—important yet peripheral to the core economic activities of firms. Early adopters of sustainable practices were frequently motivated by ethical considerations and regulatory pressures rather than a clear understanding of the economic benefits. However, as global awareness of environmental issues such as climate change, resource depletion, and social inequality has increased, the landscape of business strategy has evolved significantly.

In the early 2000s, sustainability began to gain traction as a key factor in risk management and corporate reputation. Pioneering companies recognized that sustainable practices could mitigate risks associated with environmental regulations and resource scarcity while also enhancing their brand image and customer loyalty. This realization marked the beginning of a paradigm shift—from viewing sustainability as a cost center to recognizing it as a potential source of economic value. The financial crisis of 2008 further underscored the importance of sustainable business models. As the global economy reeled from the repercussions of unsustainable financial practices, there emerged a renewed focus on long-term value creation and the need for resilient, ethical business strategies. In the aftermath, concepts such as Environmental, Social, and Governance (ESG) criteria gained prominence, offering investors new frameworks to evaluate the sustainability and ethical impact of their investments. In the present day, the integration of sustainability into business strategy is recognized not only as beneficial but essential for long-term economic success. This book aims to explore this critical relationship between sustainable business practices and economic performance, offering a comprehensive understanding of how companies can benefit financially by embedding environmental and social responsibility into their core strategies.

The volume bridges the gap between environmental and economic concerns, providing a robust argument that sustainability is a strategic economic driver. It encompasses a wide range of topics that collectively illustrate the multifaceted benefits of sustainable practices across various sectors and business functions. By exploring the rise of green finance and investments, we delve into how financial instruments and markets are adapting to support environmentally friendly projects and businesses. Government policies and initiatives are analyzed to understand how regulatory frameworks can drive sustainable business practices and innovation. Discussions on the intersection of sustainability and innovation highlight how businesses can gain a competitive edge through sustainable practices, with a focus on sectors like agriculture and technology. The introduction of conceptual models and theories provides frameworks for understanding and implementing sustainability innovations within businesses. Examining the impact of ESG objectives on investment decisions offers insights into how businesses and investors are integrating these criteria into their strategies. Exploring the synergies between advanced technologies, quality management, and supply chain strategies, we propose models for achieving sustainable development in the context of Industry 4.0. Through case studies and empirical research, effective stakeholder engagement strategies that facilitate sustainable transformations in businesses are investigated. Consumer attitudes toward sustainability, especially in the context of fintech and financial services, are analyzed using advanced analytical techniques to understand behavior and preferences. As businesses continue to navigate the complexities of the modern economic landscape, the imperative for sustainable practices becomes ever more pressing. This book provides valuable insights and practical guidance for businesses aiming to achieve economic success through sustainability. By demonstrating that sustainability can serve as a powerful economic driver, we hope to inspire a new generation of business leaders to embrace environmentally and socially responsible strategies as integral components of their corporate ethos.

We trust that this volume will contribute significantly to the ongoing discourse on sustainable business practices and their economic impact, offering readers not only theoretical insights but also practical applications and case studies from diverse industries.

Pune, India — Avichal Sharma
Ho Chi Minh City, Vietnam — Majo George
Pune, India — Pallavi Pandey
Jaipur, India — Vivek Joshi
Pune, India — Manjari Sharma

Contents

About the Editors and Contributors

About the Editors

Avichal Sharma is a Gold Medalist in MA (Economics), GATE All-India rank 172, currently pursuing Ph.D. in Subordination of Sustainable Development Goal-4 with Higher Education System in Chhattisgarh (Special reference to Bilaspur) from Durg University. He is currently working with Christ University, Pune campus as an Assistant Professor. He has around 3 years of experience in academics and has worked with renowned institutions across India. He has attended several Faculty Development Programs FDPs and workshops along with presenting papers at various international and national platforms. He has been part of over 20 international and national conferences and has even won the title of best papers for his paper.

Majo George is a Senior Lecturer, distinguished scholar, and influential educator in the Department of Business Innovation at The Business School, RMIT University Vietnam. Based at the university's Saigon South campus, Dr. George has broadened the scope of his expertise to logistics and supply chain management, circular economy, sustainability, tourism, and hospitality, and in emerging critical fields such as design thinking, fintech, cybersecurity, and fraud detection. His impressive academic credentials include a Ph.D. in Business Administration, a Master's Degree in Business Administration (MBA), a Master's Degree in Commerce (Finance and Accounting) (M.Com.), a Master's Degree in Philosophy (M.Phil.), all from India and a Graduate Diploma in Tertiary Teaching and Learning. With over 25 years of teaching experience, including more than 15 years at The Business School (TBS) at RMIT. His earlier academic positions at several renowned Vietnamese universities, including Dai Hoc Bach Khoa Ha Noi (Hanoi University of Science and Technology), the National Economic University, and Hanoi Open University, contribute to his rich teaching methodology and deep connections within the academic and business communities.

Pallavi Pandey is a Gold Medalist in MBA (Human Resource Management) from O P Jindal University, Raigarh; Masters in Commerce; NET qualified in Commerce. She is currently working with Christ University, Pune campus as an Assistant Professor, as well as the State President of the Women's Indian Chamber of Commerce and Industry. She is pursuing her Ph.D. in Management from Symbiosis International University, Pune, India, on IFRS and SDG 8.5 and 10.3, with respect to human resources. She has over 5 years of work experience in academics as well as in industry; she has worked with renowned institutions across India. As part of her entrepreneurial venture, she has been conferred with the Youngest Entrepreneur Award. She has presented more than 25 papers in seminars and conferences and attended over 15 workshops and FDPs. In 2021, she bagged the prestigious OPJEMS Scholarship which is given to only 20 management students participating from prestigious B-schools all across India.

Vivek Joshi is an Experienced Business Process Consultant, Trainer, and Passionate Academician having over 22 years of work experience. He possesses specialized skills in generating customized solutions and innovative ways to troubleshoot the issue. He focussed on building sustainable and fundamentally strong, growth-oriented ventures that would build the world of tomorrow. Currently focussing on USA, India, UK, Singapore, Middle East, and African Markets to support and handhold startups or technology-intensive ventures (pre-seed to growth stage). Small and Medium Entrepreneurs (SME) handholding is another prime area of consideration.

Manjari Sharma is an Associate Professor at Christ (Deemed to be University), Pune Lavasa Campus. She holds a Ph.D., an MBA in Finance & Marketing, M.Com., and B.Com., and specializes in Managerial Economics, Financial Management, Securities Analysis, Portfolio Management, and Operations Management. With over 15 years of teaching experience, including international experience in Maldives, Malaysia, Sri Lanka, Myanmar, and Dubai, Dr. Sharma is recognized for her innovative teaching methods and significant academic contributions. She has authored several books, including "11 Ways to Earn Passive Income: Earn Your Financial Independence" and "Innovative Pedagogy: A Beginner's Guide to Modern Teaching Strategies." Her research includes a Ph.D. thesis on the price behavior of securities in the Indian capital market. Dr. Sharma also holds a patent for an "IoT Based Fertilizer Spraying Drone." Her practical insights and dynamic approach to education have made her a respected figure in business and commerce education.

Contributors

K. Abinaya Department of Commerce (CA), KPR College of Arts Science & Research, Coimbatore, India

Deep Bhowmick Christ (Deemed to be) University, Bangalore, India

An Duong Thi Binh CIRTech Institute, HUTECH University, Ho Chi Minh City, Vietnam

Mohit Boralkar Christ (Deemed to be) University, Bangalore, India

Nilanjana Chakraborty Assam University, Silchar, India

Tuan Chu The Business School, RMIT International University, Ho Chi Minh City, Vietnam

Tien Thuy Tran Dang International School of Business (ISB), UEH University, Ho Chi Minh City, Vietnam

Duy Dang-Pham The Business School, RMIT University, Ho Chi Minh City, Vietnam

Majo George Faculty of Economics, University of Economics and Finance (UEF), Ho Chi Minh City, Vietnam

Sharad Gupta School of Business and Management, Christ (Deemed to be University), Pune, India

Irfan Ul Haq The Business School, RMIT International University, Ho Chi Minh City, Vietnam

Thinh Gia Hoang International School of Business (ISB), UEH University, Ho Chi Minh City, Vietnam

Karlheinz Kautz School of Accounting, Information Systems and Supply Chain, RMIT University, Melbourne, Australia

Varun Kesavan VIT Business School, VIT University, Vellore, India

Juthika Konwar Assam University, Silchar, India

Jyothi Kumar Christ University, Bengaluru, India

T. Mahesh Kumar Department of Commerce, Christ (Deemed to be University), Pune, India

Susnato Maji Christ University, Pune, India

Anna Maria Majo Marian College, Kuttikkanam, India

Joice Ann Mathew Christ University, Pune, India

Mai Nguyen Ngoc Hanoi Foreign Trade University, Hanoi, Vietnam

Kevin Nguyen The Business School, RMIT International University, Ho Chi Minh City, Vietnam

Son Nguyen The Business School, RMIT International University, Ho Chi Minh City, Vietnam

Thuc Hien T. Nguyen The Business School, RMIT University, Ho Chi Minh City, Vietnam

Pallavi Pandey School of Commerce, Finance and Accountancy, Christ University, Pune, India

Megha Rani Patel Department of Commerce and Financial Studies, Central University of Jharkhand, Ranchi, India

Nichapa Phraknoi The Business School, RMIT International University, Ho Chi Minh City, Vietnam

Hoang Ai Phuong The Business School, RMIT University, Ho Chi Minh City, Vietnam

Aruna Polisetty VIT Business School, VIT University, Vellore, India

Huy Truong Quang The Business School, RMIT University, Ho Chi Minh City, Vietnam

P. Radhika Christ University, Bengaluru, India

Harshal Anil Salunkhe School of Business, MIT World Peace University, Pune, India

Manjari Sharma School of Commerce, Finance and Accountancy, Christ University, Pune, India

Bateshwar Singh Department of Commerce and Financial Studies, Central University of Jharkhand, Ranchi, India

Justina Teronpi Assam University, Silchar, India

Ajay Pratap Yadav Department of Commerce and Financial Studies, Central University of Jharkhand, Ranchi, India

Towards Sustainable Development: Unveiling the Synergies of Industry 4.0, Quality Management, and Supply Chain Management Through SCQM 4.0

An Duong Thi Binh, Anna Maria Majo, Kevin Nguyen, Son Nguyen, Nichapa Phraknoi, Irfan Ul Haq, Tuan Chu, and Huy Truong Quang

Abstract Industry 4.0 adoption is a burgeoning topic in supply chain and quality management, capturing the attention of academics and professional researchers. Despite its growing prominence, there has been a dearth of assessments that simultaneously consider the interconnectedness of Industry 4.0, Quality Management, and Supply Chain Management. This gap suggests an untapped potential for exploring synergies that could foster the development of more integrated and holistic management models. This research aims to bridge this gap by developing a Supply Chain Quality Management 4.0 (SCQM 4.0) conceptual model. This model is designed to measure and implement SCQM 4.0 solutions effectively, thereby laying the groundwork for future research, particularly in experimental and applied studies. By employing a systematic literature review (SLR), this study establishes a robust knowledge base that informs the creation of the SCQM 4.0 framework. Our findings highlight the critical need for a comprehensive conceptual framework of SCQM 4.0, which is essential for advancing research in this proposed area. This framework not only seeks to enhance the understanding of the interplay between Industry 4.0 technologies and quality and supply chain management but also aims to provide practical insights for implementation, ultimately contributing to the evolution of more sophisticated and effective management practices in the Industry 4.0 era.

A. D. T. Binh
CIRTech Institute, HUTECH University, Ho Chi Minh City, Vietnam
e-mail: dtb.an@hutech.edu.vn

A. M. Majo
Marian College, Kuttikkanam, India

K. Nguyen · S. Nguyen · N. Phraknoi · I. U. Haq · T. Chu · H. T. Quang (✉)
The Business School, RMIT International University, Ho Chi Minh City, Vietnam
e-mail: kevin.nguyen@rmit.edu.vn; son.nguyen74@rmit.edu.vn; nichapa.phraknoi@rmit.edu.vn; irfan.ulhaq@rmit.edu.vn; tuan.chu2@rmit.edu.vn; huytruong.quang@rmit.edu.vn

A. Sharma et al. (eds.), *The Synergistic Effect of Sustainable Business Practices on Corporate Performance*, https://doi.org/10.1007/978-981-96-3375-3_1

1 Introduction

Supply chain quality management (SCQM), an integration of both concept management chain management and quality management, refers to the strategies a clue company uses to combine and/or integrate with the suppliers and customers for quality improvement (Da Silva et al., 2019). SCQM converts from an activity-based approach to a strategy-based approach, resulting in higher customer satisfaction through enhanced network collaboration between companies and higher execution of upstream and downstream processes for the organisation. Hence, the quality of products and services improves (Da Silva et al., 2019).

Besides, Industry 4.0 (I4.0), or the Fourth Industrial Revolution, is an emerging area with growing research and applications from researchers and industry professionals (Rocha et al., 2019). I4.0 was first introduced at the Hannover Industrial Fair in Germany in 2011. The Industrial Revolution 4.0 introduced a series of the latest technologies, merging all the physics, digital, and biology knowledge and affecting all sectors, industries, and economies (Verma & Venkatesan, 2021). The industrial revolution focuses on disruptive technologies such as artificial intelligence (AI), robotics, the Internet of Things (IoT), Biotechnology, self-driving cars, 3D printing, nanotechnology, Big Data (Big Data), etc., facilitating the creation of smart factories and supply chains, etc. (Da Silva et al., 2019).

Nevertheless, very limited assessments are conducted from the prism of all three concepts (Industry 4.0, Quality Management, and Supply Chain Management), connecting the link and exploring more synergies to support the further development of more holistic management models. Figure 1 presents the integration of the components in Industry 4.0, quality management, and supply chain management that generates the SCQM 4.0.

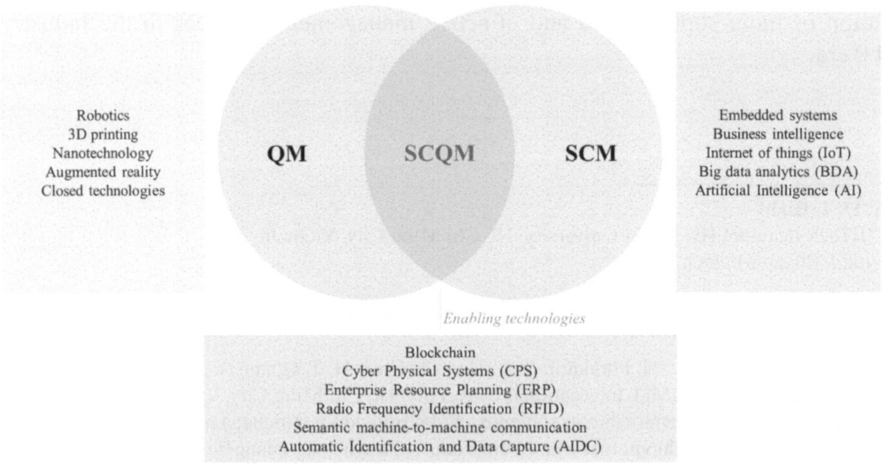

Fig. 1 Integration of Industry 4.0 of QM, SCQM, and SCM

This study aims to develop a conceptual framework for supply chain quality management 4.0 (SCQM 4.0) that can facilitate the deployment of SCQM 4.0 to drive the digital transformation of supply chain management. Systematic literature review (SLR) is applied in this study to establish a concrete knowledge base for the creation of the three-phase conceptual framework of SCQM 4.0 that integrates supply chain management and quality management under the paradigm of Industry 4.0. This study expects that the conceptual framework can be used as a "handbook" for measuring and implementing SCQM 4.0 solutions and lay the groundwork for future research in this field, especially experimental and applied studies.

The remainder of the paper is organised as follows: Sect. 2 presents the systematic literature review for obtaining relevant materials on the integration of I4.0, quality management, and supply chain management; Sect. 3 presents the results of the literature review with a discussion; Sect. 4 presents the conceptual framework of SCQM 4.0, and Sect. 5 provides the conclusion and future research directions.

2 Research Methodology

Over the years, due to a systematic literature review, an appropriate analytical model for SCQM coordination and integration has been developed. The essence of this systematic review is to aid in establishing solid knowledge bases, thereby facilitating thorough exploration and analysis. Further, SLR provides rigorous research methodology to solve research problems through clear and extensive document scanning, specific notes, importance assessments, and mapping of the "known" and "unknown" for the areas under investigation (Briner & Denyer, 2012). Insights are gained from assessments targeted at thought-provoking and future theory-building on these areas; the area of strategic management is being explored. Moreover, a systematic review is considered a powerful and testable method that initiates the development of theories, acknowledges guidelines, and draws conclusions (Ferrari, 2015). Because of the mitigating possibility of bias, the systematic review process guarantees a centralised, transparent, and repeatable assessment of required research with high confidence (Briner & Denyer, 2012). The fundamental process for a comprehensive and coherent systematic literature review is structured in three phases: planning, conducting, and reporting (Robertson et al., 2003), as presented in Table 1.

2.1 *Planning*

During the planning phase, the keywords were established to extract research topics and fundamental integration problems for these research lines. To complement the existing literature reviews and demonstrate a general view of supply chain quality management 4.0 consistent with the study objectives, in-depth research was conducted towards the discovery of uncharted territories with keyword searching, as

Table 1 Systematic Literature Review (SLR) protocol

Phase 1:	
Planning	Search Strings
	SCQM 4.0: Industry 4.0 AND supply AND chain AND quality, Digitization AND supply AND chain AND quality, Fourth Industrial revolution AND supply AND chain AND quality, Smart Manufacturing AND supply AND chain AND quality, Smart factory AND supply AND chain AND quality, Cyber Physical System AND supply AND chain AND quality, Internet of things AND supply AND chain AND quality, Industrial Internet AND supply AND chain AND quality, Big data AND supply AND chain AND quality, Blockchain AND supply AND chain AND quality.
	Literature Databases
	Scopus, Elsevier, Emerald, Wiley Online Library (Wiley), Taylor & Francis, Springer, Inderscience, Informs PubsOnline, Ebsco and ProQuest, IEEE Xplore, Web of Science, and Google Scholar
	Search Period
	1998 to June 2023
Phase 2:	
Conducting	Searching based on Planned Criteria
	Screening Identification: Identify throughout database and abstract Eligibility: Review introduction and conclusion Included: Review full text
	Analysis Descriptive analysis and thematic synthesis
Phase 3:	
Reporting	Establishing Constructs for Supply chain quality management 4.0 Concept Building Conceptual Frameworks

shown in Table 1. The journal and conference publications covered by the review will be located and extracted through the use of an aggregated database including Scopus, Elsevier, Emerald, Wiley Online Library (Wiley), Taylor & Francis, Springer, Inderscience, Informs PubsOnline, Ebsco and ProQuest, IEEE Xplore, Web of Science, and Google Scholar. Despite the overlap in the results obtained from the concurrent usage of several databases, all relevant documents are guaranteed to be collected. However, only peer-reviewed journal publications and conference proceedings were included in the review to ensure the greatest reliability in the research areas under investigation (Bui et al., 2022).

2.2 Conducting

In the implementation phase, the performance of a search step identifies appropriate articles based on the planned criteria. These articles are next vetted through the title, abstract, introduction, conclusion, and, if necessary, the entire manuscript. Different methods for synthesising and analysing qualitative data are available in the literature, including qualitative summary, qualitative meta-analysis, theoretical background, content analysis, and thematic synthesis (Barnett-Page & Thomas, 2009). Since it provides a structured approach for interpreting thematic information and facilitates the development of a holistic view of the literature under consideration, this research will focus on the thematic synthesis method. A database in MS Excel was conceived to organise, encode, and categorise the articles included in this review, grouping the studies by SCQM and SCQM 4.0 categories for descriptive analysis and summary by topic. From the publications and records on the developed database, key descriptive information, including publication date (year), database, journal, survey disciplines, and applied research methods, was collected to accumulate descriptive data.

For thematic analyses, main searches focused on proposed (for conceptual studies) and/or proven (for empirical studies) key relationships and key areas of discussion that were noted in each article in the category review (SCQM 4.0). Afterwards, the prime elements of the topics are identified, leading to the classification and coding. Ultimately, hardly any articles about SCQM 4.0 were found. Four research types were utilised in this study. Initially, the model type includes the main classification of each paper and is taken from the work of Sayama (2015) for the classification of scientific models. After defining the model type, the SCQM (coordinated and/or integrated) goals are distinguished in the second category. The third category of applied tools consists of modelling tools or techniques used in each research. Fourthly, coordination/integration mechanisms include the necessary management practices.

2.3 Reporting

The application/focus area (or topic) of each research work was extracted from the analysis of each paper. In this article, these areas of application/focus are referred to as "practice". Each practice was extracted by thoroughly reading and analysing articles with dimensions derived from various manuscript sections, such as proposals, results, conclusions, and implications. Consequently, conceptual frameworks were built.

3 Result Analysis of the SLR

3.1 Descriptive Statistic

A total of 43 articles were selected and classified as relevant by systematic literature review for the lines of research graphically illustrated in Fig. 2. In this study, Bib Excel was chosen to work on large amounts of data thanks to its versatility and compatibility with multiple computer applications, including Excel, Pajek, and Gephi (Persson et al., 2009). In addition, Bib Excel assists with preparing input data for detailed network analysis. From the results of 43/116 articles in Fig. 2, notwithstanding a more moderate quantity compared to studies on SCQM 4.0, it is indisputably fertile land for research and is of considerable interest to myriad academicians. It is predicted that the Internet of Things (IoT), Industry 4.0, supply chain, and supply chain quality are the leading keywords that help identify SCQM 4.0. Thus, based on these keywords, the proposed SCQM 4.0 will be a new research direction in this research.

Figure 3 shows that the field of SCQM4.0 has captured increasing academicians' attention since 2005. Although SCQM 4.0 first appeared in 2005, this area of research has only piqued the interest of scholars since 2014. However, compared with the business community, the growth of research in this area is probably slower. As can be seen from the chart, studies on SCQM 4.0 are likely to increase from

Fig. 2 Overview of article identification, screening, and inclusion processes

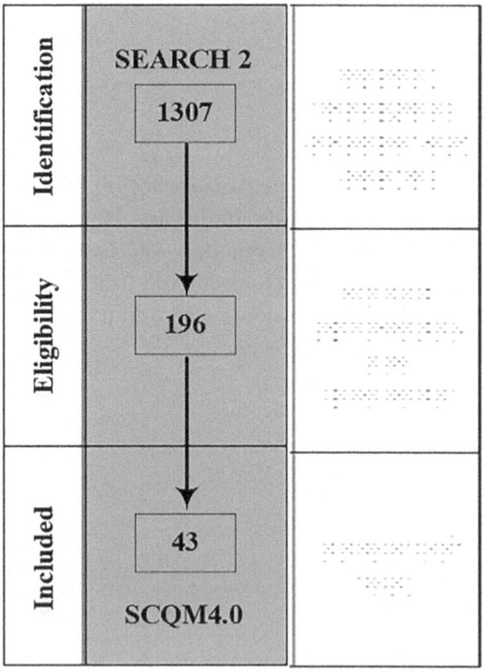

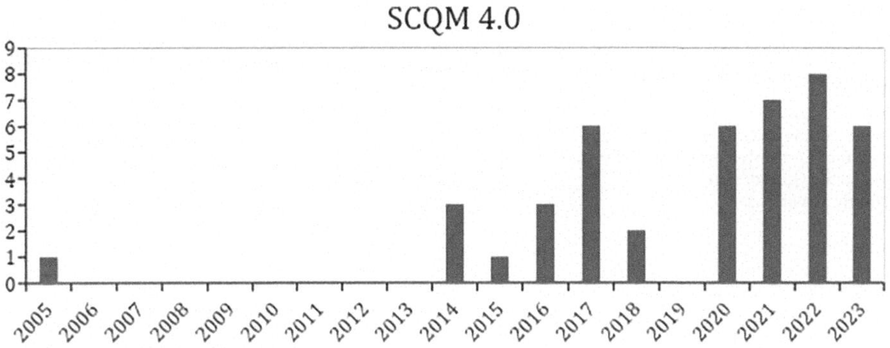

Fig. 3 Distribution of papers per year

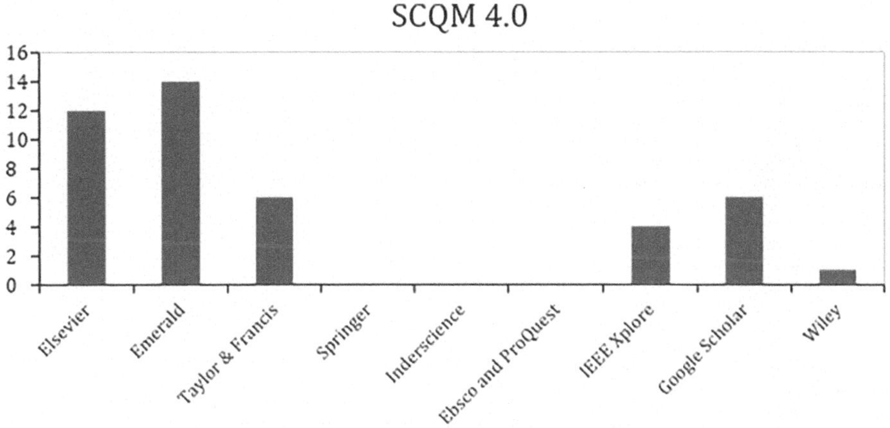

Fig. 4 Distribution of articles per database

2020 onwards, when multiple special issues have drawn attention from journals, especially international journals, such as The TQM Journal, Industrial Management & Data Systems, Food Control, International Journal of Quality & Reliability Management, and International Journal of Production Economics. Undeniably, the field of SCQM 4.0 has gained momentum in recent years. From 2014 to 2020, the number of related studies has risen twofold, especially in the Digital 4.0 research line. Considering the advantage of I4.0-related disruptive technologies for organisations from legislative bodies and customers and the needs of today's society, research topics on incorporating Engineering 4.0 into business practice are projected to surge.

Besides, Fig. 4 illustrates the number of articles per database of the antecedent literature related to SCQM. In spite of a variety of databases used in the search, the articles seem to be concentrated mainly on a few databases such as Emerald, Elsevier, and Taylor & Francis. Meanwhile, Fig. 5 describes the number of articles found in each journal for the antecedent literature related to SCQM4.0. Quality and

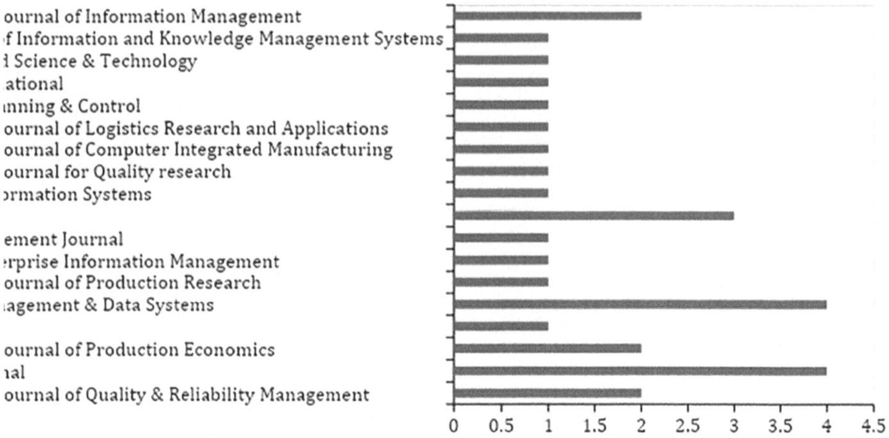

Fig. 5 Number of articles found in each journal

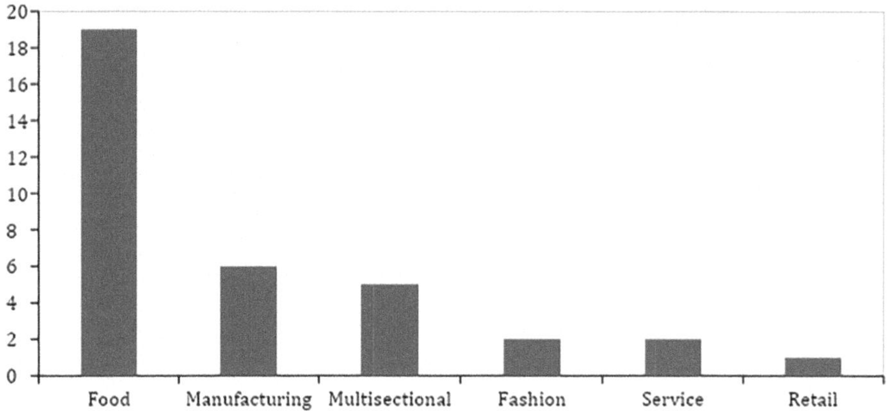

Fig. 6 Surveyed industries

supply chain-oriented journals, for example, The TQM Journal, Industrial Management & Data Systems, Food Control, International Journal of Quality & Reliability Management, and International Journal of Production Economics, have a wide range of publications on SCQM4.0.

Moreover, Fig. 6 illustrates the types of surveyed industries studied in the antecedent literature related to SCQM4.0. This finding emphasises a limited number of empirical studies that need more attention from the academic community to assess the validity and reliability of conceptual models, as recommended by Frederico et al. (2020) and Sony et al. (2020). Such empirical research primarily used data from the Food (44%), Manufacturing (14%), and Multisectional (12%) industries. The surveyed manufacturing and service industries have vastly different proportions (75% – 25%) which suggests that academics place a high value on research related to manufacturing.

Furthermore, Fig. 7 presents the types of research methodologies applied to antecedent literature related to SCQM4.0. The results show that more than 46% of the publications are empirical studies, including surveys (30%) and case studies (16%). The majority of research methodologies are applied to lend insight into how supply chain management, quality management, and digital 4.0 are implemented in various industries by administering questionnaires to companies, conducting surveys with them, or analysing case studies of firms in specific sectors. 40% of conceptual papers suggest innovative frameworks for integration, including techniques, tools, and practices. Moreover, 12% of research streams apply both qualitative and quantitative methods. Meanwhile, literature reviews contribute only 2% of the number of publications to facilitate theory-building and research directions for future directions. Carranza (2008) proposed the following two approaches: (1) knowledge-based/direction and (2) data-based/quantitative to enhance the accuracy and validity of research data results. Future studies should apply this combination of methods to showcase more understanding and bolster the confidence level of integrated research paths.

3.2 *Thematic Synthesis and Analysis*

Recent studies on SCQM4.0 are generally focused on the application of blockchain technology with the elimination of risks and the improvement of supply chain quality management. As one of the emerging and disruptive technologies, blockchain is driving the development of e-agriculture. The application of blockchain technology in various business applications emerged in supply chain management (Nguyen et al., 2022). Due to features such as transparency, censorship resistance, distributed

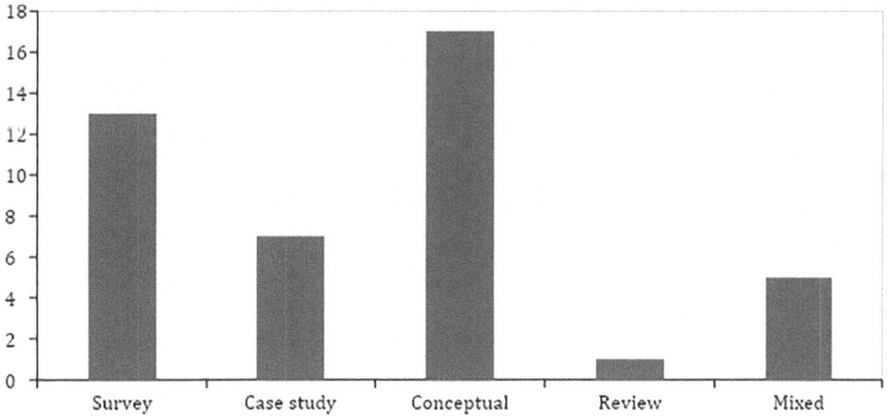

Fig. 7 Research methodologies applied in prior studies

ledger technologies (DLT), and smart contract features, the adoption of blockchain technology can improve data coordination and trust between stakeholders and construct a superior data infrastructure (Vadgama & Tasca, 2021). The adoption of blockchain technology is suggested by extant studies as a feasible application for supply chain operations management (Saurabh & Dey, 2021). Figure 8 presents a summary of the recent literature related to the applications of blockchain technology in supply chain management.

There is also a growing literature segment that proposes the integration of blockchain systems and other disruptive technologies, such as the Internet of Things (IoT), network-physical systems, and semantic machine-to-machine communication (Gupta et al., 2023, Biswas et al., 2023). The IoT is an emerging new information technology revolution that offers a paradigm shift in several areas, including QM. The IoT elevates supply chain communication to a new level by enabling seamless interaction not only between people but also between interconnected devices. It facilitates the automatic coordination of these 'things,' whether they are stored within a facility or in transit between different entities across the supply chain (Ben-Daya et al., 2019, Kemp et al., 1998). The IoT is undergoing exponential growth and gaining much attention in academic and industrial fields, providing new levels of supply chain visibility, agility, and adaptability to deal with various quality challenges (Ellis et al., 2015).

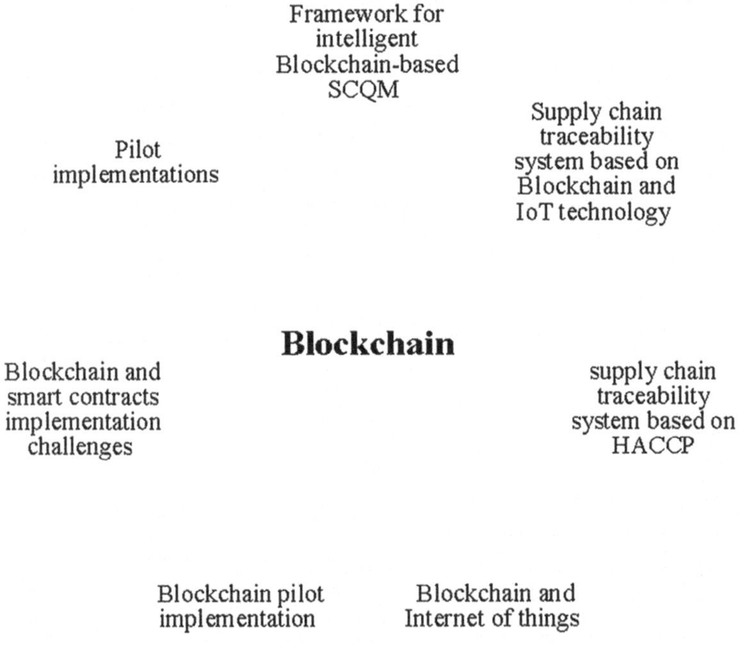

Fig. 8 Summary of recent literature on blockchain technology applications in supply chain management

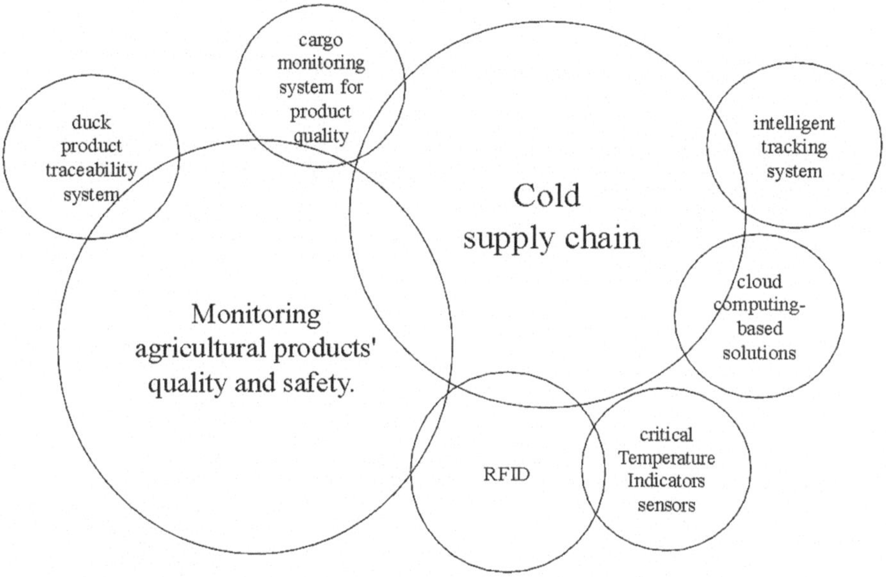

Fig. 9 Main IoT-based solutions for the supply chain

However, privacy risks and security holes remain as a consequence of the inadequacy of security technology basics (Makhdoom et al., 2019). Current security and privacy approaches do not apply to IoT due to its decentralised topology and mobile device resource constraints (Lu, 2019). There are some challenges in the large-scale deployment of IoT, such as the high investment cost of the IoT technology, the huge data volume generated by IoT, which requires robust software and hardware as well as a secure environment, and the lack of standardisation in the IoT systems that caused barriers for companies to adopt IoT (Arfi et al., 2021). Besides, the lack of standardisation leads to a case-by-case implementation approach owing to the high diversity of products and the other complicating factors stemming from insufficient government regulation and poor Internet infrastructure (Makhdoom et al., 2019). Figures 9 and 10 present a summary of the main IoT-based solutions and frameworks for the supply chain.

All of the above embarked on the findings of the SLR; we propose a definition of Supply Chain Quality Management 4.0 as below:

Supply Chain Quality Management 4.0 is a holistic and transformative approach to supply chain and quality management where companies are applying breakthrough Industry 4.0 technologies to their operations in the industry network of companies for higher quality products and services, streamlining processes and supply chain relationships to achieve high levels of customer satisfaction and benefit significant strategy for all stakeholders in the supply chain.

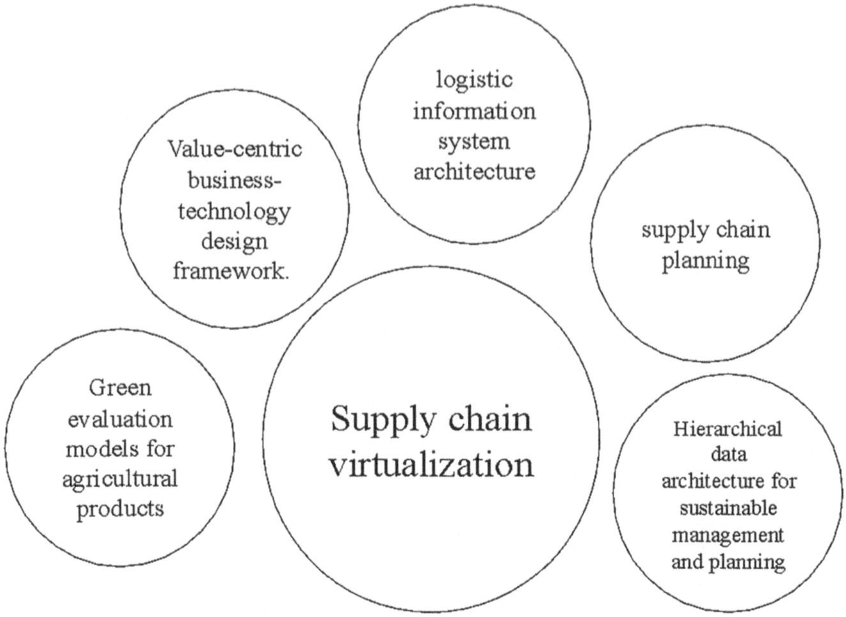

Fig. 10 Main IoT-based frameworks for the supply chain

4 The Theoretical Framework for SCQM 4.0

This paper uses SLR to generate a relevant analytical model for the coordination and integration of SCQM 4.0 over the years. The essence of this systematic review is to assist in establishing solid knowledge bases, to facilitate thorough and in-depth exploration and analysis, and to provide a rigorous methodology to solve research problems via clear and extensive document scanning, specific notes, importance assessments, and mapping of the "known" and "unknown" for the areas under investigation (Briner and Denyer, 2012). In addition to introducing the term SCQM 4.0 and proposing a definition for it, this article attempts to propose a conceptual model of SCQM 4.0 practices with the goal of accomplishing an SCQM 4.0 implementation. A conceptual framework of SCQM 4.0 includes four constructs with the corresponding characteristics, as shown in Fig. 11.

4.1 Disruptive Technologies in SCQM 4.0

Figure 12 represents the list of disruptive technologies in Supply Chain Quality Management (SCQM) 4.0 highlights the transformative impact of advanced digital tools. Research and development in Industry 4.0 (I4.0) and its associated technologies across diverse fields have significantly contributed to progressive

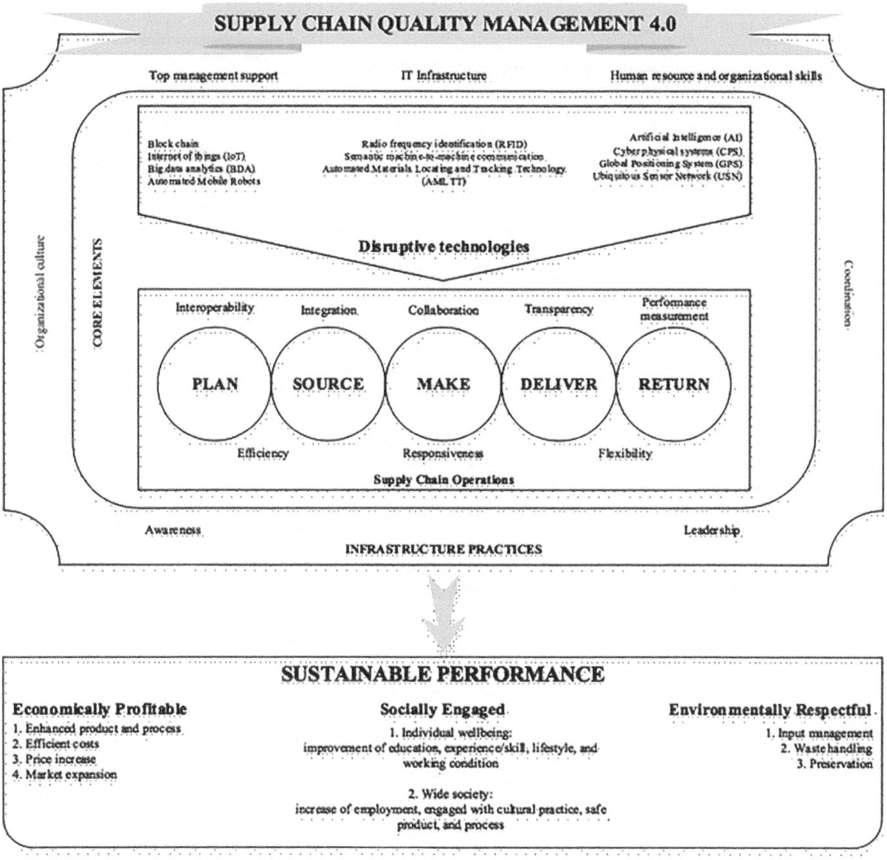

Fig. 11 SCQM 4.0 conceptual framework

social and economic development. Key initiatives have emerged since the adoption of technologies such as Blockchain, Internet of Things (IoT), Big Data Analytics (BDA), Radio Frequency Identification (RFID), Global Positioning System (GPS), Ubiquitous Sensor Network (USN), Autonomous Mobile Robots and Automated Guided Vehicles (AMLTT), semantic machine-to-machine communication, Cyber-Physical Systems (CPS), Robotics, and Artificial Intelligence (AI). These technologies have been advancing rapidly, driving innovation and efficiency. These new technologies will have implications for many business areas, including the development of new products and services, quality improvement, operations management and business models, etc., which will give rise to significant changes to the supply chain quality management (Pereira & Romero, 2017, Bienhaus & Haddud, 2018).

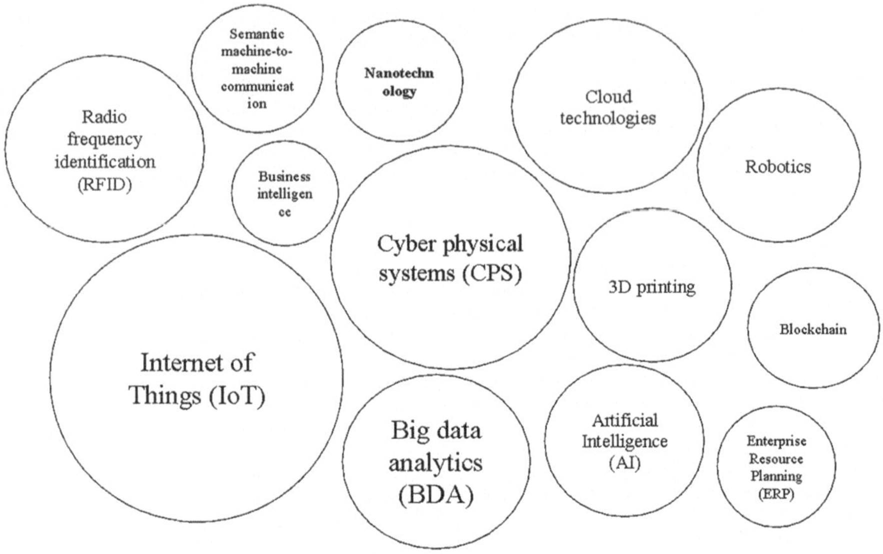

Fig. 12 List of disruptive technologies in SCQM 4.0

4.2 Supply Chain Operations

The effectiveness of these techniques will be assessed through the supply chain processes (Bui et al., 2022). To measure process performance, existing processes must be modelled. The author argues that process models describe supply chain activities through the prism of information processing. Process modelling can be considered descriptive modelling and used as an interface for developing information systems. For supply chain process modelling, there are several common approaches:

- Architecture of Integrated Information Systems (ARIS).
- Unified Modeling Language (UML).
- Integration Definition for Function Modeling (IDEF).
- Supply Chain Operations Reference (SCOR).

These models aim to describe SC processes and structures graphically in which the entities of the chain combine different planning horizons. In our study, the SCOR model was chosen because it provides the ability for elements to be divided into different hierarchical levels to facilitate detailed analysis in a structured manner. This justifies its intuitive use in various fields and explains its wide usage in both theory and industry. More importantly, this SCOR model has been recognised as the most famous model for describing supply chain processes and measuring their performance (acceleration).

The SCOR model identifies best practices, performance metrics, and software functional requirements for each core process, sub-process, and supply chain operation. The SCOR model offers a foundational structure and standard terminology to help companies unify a variety of management tools, such as business process reengineering, benchmarking, and practice analysis. In addition, it contains a description of the planning, sourcing, generation, delivery, and return processes in SC at different levels of abstraction. The most vital indicators of supply chain performance are identified and classified. Level 1 metrics are evaluated using the best-worst method, and their internal levels are evaluated using the Pythagorean Analytic Hierarchy Process (AHP) method. The weighted significance of each level 2 and level 3 metric is obtained. Both the public and private sectors can use the SCOR model to strengthen their supply chain strategies in the global world.

As aforementioned, the potential outcomes of SCOR processes are the driving force behind disruptive technologies. To exemplify, the adoption of disruptive technologies results in two issues: (1) Interoperability between technologies and (2) cooperation between technologies and systems. This interoperability and collaboration will significantly impact the SCOR process's efficiency and effectiveness, facilitating integration between supply chain actors. In vertical integration, techniques such as Enterprise Resource Planning (ERP) and Systems Applications and Products (SAP) are the key to integration between functions in the company. Horizontal integration throughout the supply chain can be achieved through Big Data Analytics, Cloud Technology, the Internet of Things, etc. (Fig. 13).

The SCOR model is one of the best models for depicting supply chain operations in operations management for research and practice. There are radical changes in supply chain structure as well as technological developments in today's information age. Companies are likely to develop and manage an efficient supply chain structure through SCOR.

4.3 Infrastructure Practices

The SCQM 4.0 uses many breakthrough technologies to upgrade product quality in the supply chain, which requires the availability of IT infrastructures for the development, initial deployment and management, and ongoing strategic evolution of disruptive technologies. To deploy technologies effectively, employees need to receive ongoing training from entities. The required SCQM 4.0 skills include technical aspects and transferable skills such as adaptability, critical thinking, creativity, teamwork, and knowledge exchange (Schallock et al., 2018). Therefore, more than a transformational leadership style, it should also be characterised by considering learning and innovation. However, current transformational leadership styles are limited to idealising impact, inspiring motivation, intellectual stimulation, and providing foresight (Bass, 1985). The lack of standardisation in IoT systems can lead to difficulty connecting between different systems, platforms, and standards. Indeed, there is a need for fast detection sensors that are resistant to harsh environments (Ping et al., 2018).

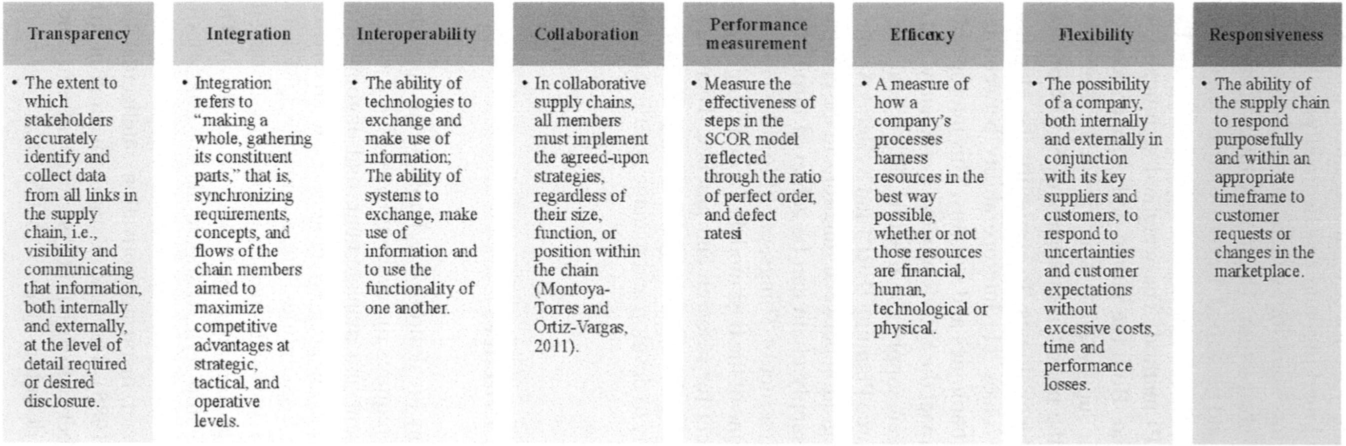

Fig. 13 SCOR indicators

Besides, the resistance of stakeholders to change during the implementation of the technologies in the organisations and business processes hinders the technological adoption and results in challenges for the SCQM 4.0 implementation; hence, the commitment from senior leadership to facilitate these changes is pivotal. Moreover, the success of SCQM 4.0 also depends on incorporating technologies into business strategies; thus, top management support and knowledge-oriented leadership are demanded. The lack of standardisation leads to a case-by-case implementation approach due to the high product diversity (Badia-Melis et al., 2018). Finally, other complicating factors stem from insufficient government regulation and poor Internet infrastructure (Kamble et al., 2019). Figure 14 describes SCQM4.0 infrastructure practices.

4.4 Sustainable Performance in SCQM 4.0

Sustainability has emerged as a critical concern for businesses worldwide in recent years. Beyond ethical considerations, organisations increasingly recognise the economic benefits of sustainable performance (Ghobakhloo et al., 2023). This study delves into the three dimensions of sustainability and their components: (1) economically profitable, (2) socially engaged, and (3) environmentally respectful.

Top management support	Top management is supposed to shape a proper strategy which ensures the organization's purposes are aligned with the implementation of new technology.
IT Infrastructure	IT capabilities and resources should be made available for the initial development, implementation and ongoing management and evolution of disruptive technologies.
Human resource and organisational skills	Management structure, HR strategy, work environment and skill development for Supply Chain Quality Management 4.0, especially for the adoption of new technologies
Coordination	Coordination refers to active and direct cooperation, and it is obtained by sending the correct signals or sharing the correct information and the same policies (Moharana et al. 2012)
Organisational culture	A set of norms, beliefs and values shared by members of the organisation towards Supply Chain Quality Management 4.0 understanding.
Awareness	A clear understanding among all entities in the supply chain regarding the benefits and requirements of the Supply Chain Quality Management 4.0
Leadership	A clear understanding of the evolutionary nature and strategic implications of Supply Chain Quality Management 4.0 to make the right decisions with regards to budget and resource allocations

Fig. 14 Description of infrastructure practices

4.4.1 Economically Profitable

We investigate the indicators related to Enhanced Product and Process, efficient costs, price increase, and market expansion. We delve into the notion of economically profitable performance by shedding light on how companies can align their economic interests with environmental and social responsibility.

(1) Enhanced Product and Process. One significant indicator of the economic profitability of sustainable performance is the enhancement of products and processes (Tseng et al., 2021). By incorporating sustainable design principles, companies can develop products that appeal to environmentally conscious consumers (Baines et al., 2012).
(2) Efficient Costs. Sustainable practices can lead to cost efficiencies throughout a company's operations. Adopting energy-efficient technologies, optimising supply chain management, and reducing waste can result in substantial savings (Bocken et al., 2016).
(3) Price Increase. Contrary to conventional assumptions, sustainable products often command premium prices. Consumers are increasingly willing to pay a higher price for products that align with their values, such as those that are eco-friendly, socially responsible, or ethically sourced (Bocken et al., 2015).
(4) Market Expansion. Sustainable performance can facilitate market expansion by unlocking new customer segments and business opportunities. As sustainability gains prominence in consumer preferences, businesses that prioritise sustainability can tap into niche markets and attract new customers (Soni & Kodali, 2016, Lim & Sonko, 2019).

4.4.2 Socially Engaged

Sustainable performance goes beyond social dimensions that are integral to its long-term viability (Bocken et al., 2015). We investigate the indicators related to individual well-being and society as a whole. We delve into the notion of socially engaged sustainable performance by underscoring the importance of social responsibility in sustainable practices and demonstrate how they can contribute to societal progress.

(1) Individual Well-being. The first set of indicators focuses on individual well-being, as sustainable performance can directly influence the lives of individuals within organisations and communities (Çıkmak & Kesici, 2023). It comprises the following sub-components:

    Improvement of Education. Sustainable practices often involve investments in employee education and skill development. By providing training and educational opportunities, companies contribute to their employee's personal growth and professional advancement (Ijaz Baig & Yadegaridehkordi, 2023).

Experience/Skill Enhancement: Sustainable performance encourages organisations to provide meaningful work experiences that empower individuals and foster personal development (Talib et al., 2020).
Lifestyle Enhancement: Sustainable practices can contribute to improving individuals' lifestyles by promoting work–life balance, supporting employee wellness programmes, and fostering a positive organisational culture (Tseng et al., 2022).
Working Conditions: Sustainable performance includes efforts to ensure safe, healthy, and fair working conditions. By prioritising employee safety, fair wages, and work–life balance, organisations foster an environment that values and supports the well-being of their workforce (Ijaz Baig & Yadegaridehkordi, 2023).

(2) Wide Society. Sustainable performance extends its positive impact beyond individual well-being to wider societal benefits (Bocken et al., 2015).

Increase in Employment: Sustainable businesses often create new employment opportunities by investing in renewable energy, environmental conservation, and socially responsible initiatives (Naeemah & Wong, 2022).
Engagement with Cultural Practices: Socially engaged sustainable performance involves respecting and engaging with diverse cultural practices (Bocken et al., 2016).
Safe Product and Process: Sustainable performance emphasises the production of safe products and processes that prioritise consumer health and safety (Bocken et al., 2016).

4.4.3 Environmentally Respectful

Environmental respect and responsible resource management are fundamental aspects of sustainable performance (Gould & Colwill, 2015). We investigate the indicators related to input management, waste handling, and preservation, emphasising the significance of environmentally respectful practices. By examining these indicators, we aim to highlight the positive impact businesses can have on the environment and underscore the importance of integrating environmental considerations into sustainable performance strategies.

(1) Input Management. Effective input management is a key indicator of environmentally respectful sustainable performance, encompassing the responsible use and conservation of resources throughout a company's operations (Tseng et al., 2021).

Resource Efficiency: Sustainable businesses prioritise efficient use of resources, such as energy, water, and raw materials, which can reduce resource consumption, optimise production processes, and minimise waste; companies can achieve significant environmental benefits and cost savings (Alnajem et al., 2021).

Sustainable Sourcing: Adopting sustainable sourcing practices involves selecting suppliers who adhere to responsible environmental standards (Beiki et al., 2021).
Life Cycle Assessment: Businesses can conduct life cycle assessments to evaluate the environmental impact of their products or services throughout their entire life cycle, from raw material extraction to disposal (Bocken et al., 2016).

(2) Waste Handling. Efficient waste handling is crucial for environmentally respectful, sustainable performance, as it minimises pollution, conserves resources, and promotes circular economy principles (Araee et al., 2020).

Waste Reduction: Sustainable businesses implement strategies to minimise waste generation through process optimisation, product redesign, and material substitution (Sadeghi Ahangar et al., 2021).
Recycling and Reuse: Encouraging recycling and reuse initiatives helps divert waste from landfills and conserve valuable resources (Bocken et al., 2016).
Hazardous Waste Management: Environmentally respectful sustainable performance involves proper handling and disposal of hazardous materials to prevent pollution and protect ecosystems and human health (Araee et al., 2020).

(3) Preservation. The preservation of natural resources and ecosystems is a critical aspect of environmentally respectful, sustainable performance (Bocken et al., 2016).

Biodiversity Conservation: Sustainable businesses prioritise the protection and restoration of biodiversity by minimising habitat destruction, supporting conservation efforts, and implementing sustainable land management practices (Bruel et al., 2019).
Environmental Stewardship: Engaging in environmental stewardship involves actively monitoring and mitigating negative environmental impacts caused by business activities (Tseng et al., 2023).
Climate Change Mitigation: Sustainable performance requires businesses to actively contribute to global efforts to mitigate climate change (Bocken et al., 2016).

5 Conclusions and Future Research

The advent of I4.0 highlights the I4.0 disruption technologies that enable SCQM to bolster quality and performance across supply chain operations. SCQM 4.0 joins manufacturing with disruptive technologies that can maximise outputs and create high-quality products and services whilst utilising minimum resources (Sony & Naik, 2019). This study is theoretically and practically significant. In practical terms, the conceptual frameworks proposed can effectively support practitioners participating in SCQM 4.0 implementation programmes. The SCQM 4.0 conceptual framework provides a complete overview of all the aspects that need meticulous consideration in order to achieve successful implementation and management. The advantages and limitations will be pointed out and analysed to provide clear levels of progression underpinned by the structures developed. The SCQM 4.0 framework can be regarded as a "guideline" assisting practitioners in achieving excellence in the implementation of SCQM 4.0. In other words, the research aims to develop conceptual models, which can be used as a "handbook" for measuring and implementing SCQM 4.0 solutions, as well as creating premises for further research followed in this field, especially experimental/applied studies.

From the search results of 43 articles in Fig. 2, it is noticeable that there is a limited amount of research revolving around SCQM 4.0. This emerging area of research is expected to expand on the strengths, synergies, and established relationships between quality, supply chain, and digital 4.0, contributing towards a quality supply chain and pioneer. Under the SCQM 4.0 conceptual framework, a clear development path which is underpinned by the SCQM 4.0 architecture is demonstrated. Business owners are encouraged to perform gap analysis and direct strategic implementation to deliver an excellent SCQM 4.0. The proposed SCQM 4.0 frameworks make relevant and significant contributions to the academic community. Subsequent exploratory qualitative studies should be carried out in different organisations and industries to explore any additional key structures specific to an area of identification of SCQM 4.0 keys for the organisation and industry selection.

References

Alnajem, M., Mostafa, M. M., & ElMelegy, A. R. (2021). Mapping the first decade of circular economy research: A bibliometric network analysis. *Journal of Industrial and Production Engineering, 38*(1), 29–50.

Araee, E., Manavizadeh, N., & Aghamohammadi Bosjin, S. (2020). Designing a multi-objective model for a hazardous waste routing problem considering flexibility of routes and social effects. *Journal of Industrial and Production Engineering, 37*(1), 33–45.

Arfi, W. B., Nasr, I. B., Khvatova, T., et al. (2021). Understanding acceptance of eHealthcare by IoT natives and IoT immigrants: An integrated model of UTAUT, perceived risk, and financial cost. *Technological Forecasting and Social Change, 163*, 120437. https://doi.org/10.1016/j.techfore.2020.120437

Badia-Melis, R., Mc Carthy, U., Ruiz-Garcia, L., et al. (2018). New trends in cold chain monitoring applications-a review. *Food Control, 86*, 170–182.

Baines, T., Brown, S., Benedettini, O., et al. (2012). Examining green production and its role within the competitive strategy of manufacturers. *Journal of Industrial Engineering and Management, 5*, 53–87.

Barnett-Page, E., & Thomas, J. (2009). Methods for the synthesis of qualitative research: A critical review. *BMC Medical Research Methodology, 9*(1), 1–11.

Bass, B. M. (1985). Leadership: Good, better, best. *Organizational Dynamics, 13*(3), 26–40.

Beiki, H., Mohammad Seyedhosseini, S. V., Ponkratov, V., et al. (2021). Addressing a sustainable supplier selection and order allocation problem by an integrated approach: A case of automobile manufacturing. *Journal of Industrial and Production Engineering, 38*(4), 239–253.

Ben-Daya, M., Hassini, E., & Bahroun, Z. (2019). Internet of things and supply chain management: A literature review. *International Journal of Production Research, 57*(15–16), 4719–4742.

Bienhaus, F., & Haddud, A. (2018). Procurement 4.0: Factors influencing the digitisation of procurement and supply chains. *Business Process Management Journal, 24*(4), 965–984. https://doi.org/10.1108/BPMJ-06-2017-0139

Biswas, D., Jalali, H., Ansaripoor, A. H., et al. (2023). Traceability vs. sustainability in supply chains: The implications of blockchain. *European Journal of Operational Research, 305*(1), 128–147. https://doi.org/10.1016/j.ejor.2022.05.034

Bocken, N. M., De Pauw, I., Bakker, C., et al. (2016). Product design and business model strategies for a circular economy. *Journal of Industrial and Production Engineering, 33*(5), 308–320.

Bocken, N., Rana, P., & Short, S. W. (2015). Value mapping for sustainable business thinking. *Journal of Industrial and Production Engineering, 32*(1), 67–81.

Briner, R. B., & Denyer, D. (2012). Systematic review and evidence synthesis as a practice and scholarship tool. In *Handbook of evidence-based management: Companies, classrooms and research* (pp. 112–129).

Bruel, A., Kronenberg, J., Troussier, N., et al. (2019). Linking industrial ecology and ecological economics: A theoretical and empirical foundation for the circular economy. *Journal of Industrial Ecology, 23*(1), 12–21.

Bui, L. T. C., Carvalho, M., Pham, H. T., et al. (2022). Supply chain quality management 4.0: Conceptual and maturity frameworks. *International Journal of Quality & Reliability Management (ahead-of-print).*

Carranza, E. J. M. (2008). *Geochemical anomaly and mineral prospectivity mapping in GIS*. Elsevier.

Çıkmak, S., & Kesici, B. (2023). Analysis of barriers to the adoption of circular supply chain management: A case study in the air conditioning industry. *Journal of Industrial and Production Engineering, 40*(4), 287–300.

Da Silva, V. L., Kovaleski, J. L., & Pagani, R. N. (2019). Technology transfer in the supply chain oriented to industry 4.0: A literature review. *Technology Analysis & Strategic Management, 31*(5), 546–562.

Ellis, S., Morris, H. D., & Santagate, J. (2015). IoT-enabled analytic applications revolutionize supply chain planning and execution. *International Data Corporation (IDC) White Paper, 13*, 259697.

Ferrari, R. (2015). Writing narrative style literature reviews. *Medical Writing, 24*(4), 230–235.

Frederico, G. F., Garza-Reyes, J. A., Anosike, A., et al. (2020). Supply chain 4.0: Concepts, maturity and research agenda. *Supply Chain Management: An International Journal, 25*(2), 262–282.

Ghobakhloo, M., Iranmanesh, M., Tseng, M.-L., et al. (2023). Behind the definition of industry 5.0: A systematic review of technologies, principles, components, and values. *Journal of Industrial and Production Engineering, 40*, 1–16.

Gould, O., & Colwill, J. (2015). A framework for material flow assessment in manufacturing systems. *Journal of Industrial and Production Engineering, 32*(1), 55–66.

Gupta, S., Modgil, S., Choi, T.-M., et al. (2023). Influences of artificial intelligence and blockchain technology on financial resilience of supply chains. *International Journal of Production Economics, 261*, 108868. https://doi.org/10.1016/j.ijpe.2023.108868

Ijaz Baig, M., & Yadegaridehkordi, E. (2023). Exploring moderating effects of industry 4.0 adoption on sustainable performance of Malaysian manufacturing organizations. *Journal of Industrial and Production Engineering, 40*(4), 271–286.

Kamble, S. S., Gunasekaran, A., Parekh, H., et al. (2019). Modeling the internet of things adoption barriers in food retail supply chains. *Journal of Retailing and Consumer Services, 48*, 154–168.

Kemp, R., Schot, J., & Hoogma, R. (1998). Regime shifts to sustainability through processes of niche formation: The approach of strategic niche management. *Technology Analysis & Strategic Management, 10*(2), 175–198.

Lim, S. S., & Sonko, L. K. (2019). Linking corporate sustainability and innovation in supply chain management–evidence of a Taiwan leading glass recycling company. *Technology Analysis & Strategic Management, 31*(8), 957–971.

Lu, Y. (2019). The blockchain: State-of-the-art and research challenges. *Journal of Industrial Information Integration, 15*, 80–90. https://doi.org/10.1016/j.jii.2019.04.002

Makhdoom, I., Abolhasan, M., Abbas, H., et al. (2019). Blockchain's adoption in IoT: The challenges, and a way forward. *Journal of Network and Computer Applications, 125*, 251–279. https://doi.org/10.1016/j.jnca.2018.10.019

Naeemah, A. J., & Wong, K. Y. (2022). Positive impacts of lean manufacturing tools on sustainability aspects: A systematic review. *Journal of Industrial and Production Engineering, 39*(7), 552–571.

Nguyen, A. H., Hoang, T. G., Nguyen, L. Q. T., et al. (2022). Design thinking-based data analytic lifecycle for improving management control in banks. *Technology Analysis & Strategic Management, 36*, 1–16.

Pereira, A. C., & Romero, F. (2017). A review of the meanings and the implications of the Industry 4.0 concept. *Procedia Manufacturing, 13*, 1206–1214.

Persson, O., Danell, R., & Schneider, J. W. (2009). How to use Bibexcel for various types of bibliometric analysis. In *Celebrating scholarly communication studies: A Festschrift for Olle Persson at his 60th Birthday* (Vol. 5, pp. 9–24).

Ping, H., Wang, J., Ma, Z., et al. (2018). Mini-review of application of IoT technology in monitoring agricultural products quality and safety. *International Journal of Agricultural and Biological Engineering, 11*(5), 35–45.

Robertson, G. R., Field, J., Goodwin, B., et al. (2003). Transgenic mouse models of human CYP3A4 gene regulation. *Molecular Pharmacology, 64*(1), 42–50.

Rocha, C. F., Mamédio, D. F., & Quandt, C. O. (2019). RETRACTED ARTICLE: Startups and the innovation ecosystem in industry 4.0. *Technology Analysis & Strategic Management, 31*(12), 1474–1487.

Sadeghi Ahangar, S., Sadati, A., & Rabbani, M. (2021). Sustainable design of a municipal solid waste management system in an integrated closed-loop supply chain network using a fuzzy approach: A case study. *Journal of Industrial and Production Engineering, 38*(5), 323–340.

Saurabh, S., & Dey, K. (2021). Blockchain technology adoption, architecture, and sustainable agri-food supply chains. *Journal of Cleaner Production, 284*, 124731. https://doi.org/10.1016/j.jclepro.2020.124731

Sayama, H. (2015). *Introduction to the modeling and analysis of complex systems*. Open SUNY Textbooks.

Schallock, B., Rybski, C., Jochem, R., et al. (2018). Learning factory for industry 4.0 to provide future skills beyond technical training. *Procedia Manufacturing, 23*, 27–32.

Soni, G., & Kodali, R. (2016). Interpretive structural modeling and path analysis for proposed framework of lean supply chain in Indian manufacturing industry. *Journal of Industrial and Production Engineering, 33*(8), 501–515.

Sony, M., Antony, J., & Douglas, J. A. (2020). Essential ingredients for the implementation of quality 4.0: A narrative review of literature and future directions for research. *The TQM Journal, 32*(4), 779–793.

Sony, M., & Naik, S. S. (2019). Ten lessons for managers while implementing industry 4.0. *IEEE Engineering Management Review, 47*(2), 45–52.

Talib, F., Asjad, M., Attri, R., et al. (2020). A road map for the implementation of integrated JIT-lean practices in Indian manufacturing industries using the best-worst method approach. *Journal of Industrial and Production Engineering, 37*(6), 275–291.

Tseng, M.-L., Li, S.-X., Lin, C.-W. R., et al. (2023). Validating green building social sustainability indicators in China using the fuzzy Delphi method. *Journal of Industrial and Production Engineering, 40*(1), 35–53.

Tseng, M.-L., Lim, M. K., Helmi Ali, M., et al. (2022). Assessing the sustainable food system in Thailand under uncertainties: Governance, distribution and storage drive technological innovation. *Journal of Industrial and Production Engineering, 39*(1), 1–18.

Tseng, M.-L., Tran, T. P. T., Ha, H. M., et al. (2021). Sustainable industrial and operation engineering trends and challenges toward industry 4.0: A data driven analysis. *Journal of Industrial and Production Engineering, 38*(8), 581–598.

Vadgama, N., & Tasca, P. (2021). An analysis of Blockchain adoption in supply chains between 2010 and 2020. *Frontiers in Blockchain, 4*, 610476. https://doi.org/10.3389/fbloc.2021.610476

Verma, A., & Venkatesan, M. (2021). Industry 4.0 workforce implications and strategies for organisational effectiveness in Indian automotive industry: A review. *Technology Analysis & Strategic Management, 35*(10), 1–9.

A Conceptual Framework to Navigate Sustainability Innovation Implementation: The Duet of Dynamic Capabilities Theory and Institutional Theory

Thuc Hien T. Nguyen, Karlheinz Kautz, Hoang Ai Phuong, Duy Dang-Pham, Majo George, and Huy Truong Quang

Abstract Sustainability has been put on top of the list by almost all stakeholders in our society due to the alarming climate and social issues around the world. Therefore, managers have to bring sustainability into the core of business innovation to ensure their survival, avoid reputation and legitimacy risks as well as seize competitive advantages. This chapter proposes a comprehensive framework to guide organisations to implement sustainability innovation to gain competitive advantages. Dynamic capabilities theory, neo-institutional theory, and institutional logics perspective are employed to provide a comprehensive framework to understand, analyse, and implement sustainability innovation. With the rapid development of technologies, the landscape for sustainability practices is also evolving at a tremendous pace. In this setting, recent literature has promoted dynamic capabilities theory as a potential framework for firms to understand how they can adjust their capabilities for the successful execution of sustainability innovation. However, some researchers criticised dynamic capabilities theory for ignoring contextual factors and environmental pressures which are imperative in shaping the values of social obligations that organisations have to commit to. In this conceptual research, the neo-institutional theory and institutional logics perspective are put in place to account for the institutional contextual factors that impact sustainability innovation implementation of organisations. In extant research, these theories are often applied

T. H. T. Nguyen · H. A. Phuong · D. Dang-Pham (✉) · H. T. Quang
The Business School, RMIT University, Ho Chi Minh City, Vietnam
e-mail: s3979937@student.rmit.edu.au; phuong.hoangai@rmit.edu.vn; duy.dangphamthien@rmit.edu.vn; huytruong.quang@rmit.edu.vn

K. Kautz
School of Accounting, Information Systems and Supply Chain, RMIT University, Melbourne, Australia
e-mail: karlheinz.kautz@rmit.edu.au

M. George
Faculty of Economics, University of Economics and Finance (UEF), Ho Chi Minh City, Vietnam

A. Sharma et al. (eds.), *The Synergistic Effect of Sustainable Business Practices on Corporate Performance*, https://doi.org/10.1007/978-981-96-3375-3_2

separately which might impede their power in enabling understanding of the innovation implementation. This conceptual research integrates these theories and perspective so that they can enrich each other to provide a holistic understanding of sustainability innovation execution.

1 Introduction

In the context of continuous catastrophic natural disasters caused by climate change and urgent social issues happening around the world, both external and internal stakeholders of firms are increasingly aware of sustainability practices, which is a strong driving force for firms to engage in sustainability development (Nguyen et al., 2023). By committing to sustainability practices, firms can secure their competitive advantages and legitimacy as well as perform their ecological responsibility (Bansal & Roth, 2000). This growing trend has firmly driven investors' attention towards the triple bottom line of economic, social, and environmental performance (Obermann & Velte, 2018). The ability to earn profit while still satisfying or exceeding social expectations about corporate sustainability performance can bring firms vital competitive advantages (Vilanova et al., 2009; Martinuzzi & Krumay, 2013).

With this aim at the forefront of their business strategy, it is indispensable for companies to understand how to implement their sustainability innovations in a way that enables them to secure economic gains while achieving their social and environmental performance goals. To achieve environmental and social performance targets, innovation is often employed as it can bring about product and service renewal and organisational change towards sustainability (Adams et al., 2016). Although the literature on sustainability innovations has grown tremendously for the last decade (Harsanto et al., 2024), there is a paucity of sound theoretical grounding in research that develops models for firms to reach more advanced levels of sustainability innovations (Dangelico et al., 2017). To bridge this gap in the extant literature, this chapter develops a comprehensive framework for firms to navigate the implementation of sustainability innovations by integrating dynamic capabilities (DCs) theory, neo-institutional theory, and the institutional logics perspective.

There has been growing interest in employing DCs theory (Teece et al., 1997) as a theoretical framework to investigate sustainability innovation management (Darmani et al., 2017; Dangelico, 2016; Amui et al., 2017). Sustainability innovation requires deliberate adjustments to organisational philosophy and values, products, processes, or practices to make environmental and social contributions along with economic benefits (Adams et al., 2016). Scholars have increasingly regarded DCs theory as a suitable framework for studying these changes as the focal of DCs theory is about how firms can integrate, build, and reconfigure their internal and external resources to maintain inimitable competitive advantages in a fast-changing environment (Teece, 2007; Teece et al., 1997). However, DCs theory has been often criticised for neglecting the contextual condition which might impact DCs themselves as well as the entities as a whole (Barreto, 2010). This might leave out essential factors that can shape how organisations operate and execute their sustainability innovations.

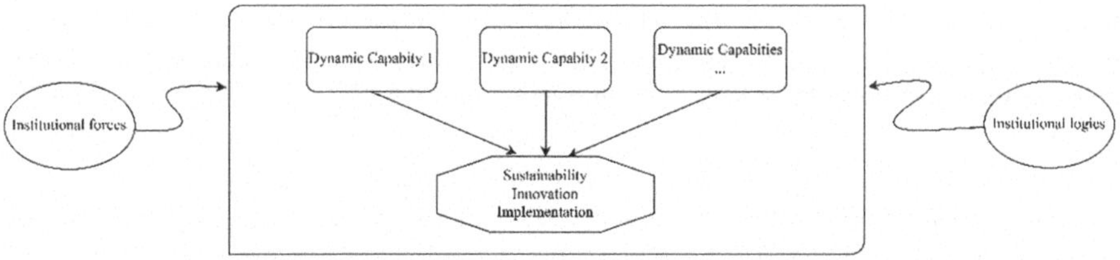

Fig. 1 Initial conceptual framework

To address these critics, this chapter employs neo-institutional theory (DiMaggio & Powell, 1983) and institutional logics perspective (Friedland & Alford, 1991; Thornton & Ocasio, 2008; Thornton et al., 2012) in combination with DCs theory to create a theoretical framework for firms to steer their journey in implementing sustainability innovations. On the other hand, DC theory also addresses issues that are not covered in neo-institutional theory and institutional logics perspective. The critiques on institutional theory are mostly about its overfocus on stability and homogeneity among organisations which undermines firms' capability to innovate and secure competitive advantages (Deephouse et al., 2017; Suddaby & Viale, 2011; Greenwood et al., 2017). This is the point where DCs theory can support the framework by providing the notion about companies' abilities to revolutionise themselves to gain competitive advantages. Fig. 1 presents the initial conceptual framework of this chapter.

The following section discusses the roles of DCs in sustainability innovation implementation. This is followed by the section about the roles of institutional forces and institutional logics at different levels during sustainability innovation implementation and how these factors impact the formation and functions of DCs. This chapter is finalised with a conceptual framework for firms to navigate their sustainability innovation execution.

2 Dynamic Capabilities Theory as a Framework for Sustainability Innovation Implementation

2.1 Brief Introduction About Dynamic Capabilities Theory

The dynamic capabilities theory has been regarded by scholars as a propitious framework for studying sustainability innovation management (Darmani et al., 2017; Amui et al., 2017). Implementation of sustainability innovations requires organisations to take into account ecological and social considerations in addition to the traditional economic concerns in business activities (Hall & Vredenburg, 2004). The shift in the strategic business approach, along with fast-changing technologies, creates a dynamic landscape for organisations to compete as they now confront major challenges in transforming their product design and marketing practices (Sargut & McGrath, 2011) and organisational operational processes (Mousavi et al., 2018). Therefore, DCs theory is considered an appropriate approach to understanding sustainability innovation as this is a dominant framework for studying strategic organisational changes (Ellström et al., 2022; Schilke et al., 2018). DCs theory focuses on how firms can integrate, build, and reconfigure their internal and external resources to conceive and maintain inimitable competitive advantages in a rapidly changing environment (Teece, 2007; Teece et al., 1997).

Among various approaches to DCs in extant literature, the most influential is the procedural typology developed by Teece (2007). Teece (2007) disaggregated DCs into three sequential clusters of higher-order DCs, including sensing, seizing, and transforming capabilities. Teece (2007) explicated the nature of DCs as abilities, capabilities, processes, and routines. To thoroughly understand about DCs, we must investigate the concept in the overall context of the organisational capabilities portfolio which can be divided into two levels (Winter, 2003). The basic level is the level of ordinary capabilities which are procedures and routines that firms follow to "achieve technical efficiency and 'doing things right' in business functions of operations, administration and governance" (Teece, 2014). These ordinary capabilities enable firms to earn short-term profits. On the next level, dynamic capabilities can be categorised into "microfoundations" and higher-order capabilities (Teece, 2007). In his seminal work, Teece (2007) explained that "microfoundations" are the unique skills, organisational structures, processes, decision rules, and disciplines that adjust and recombine ordinary capabilities. For example, Eisenhardt and Martin (2000) proposed several critical microfoundations of DCs; these microfoundations included cross-functional R&D teams, new product development routines, quality control routines, technology transfer and/or knowledge transfer routines, and certain performance measurement systems. These microfoundations are guided by higher-order DCs. These higher-order DCs are introduced by Teece (2007) in a framework of three sequential clusters: sensing, seizing, and transforming. Firstly, sensing DCs are concerned with scanning, learning, and interpreting (Teece, 2007) opportunities and threats. These opportunities and threats might arise from changes in customers' demand, new technologies, or activities of suppliers and competitors. After identifying opportunities and threats, in the seizing phase, firms need to adjust and reconfigure their existing resources to respond to those opportunities and threats to capture values from them. In particular, this phase involves refining or transforming processes and even business models to respond to the opportunities and threats sensed. Finally, transforming capabilities are the capabilities to continuously reconfigure organisational resources and structures during the continuous growth of firms and the changing of the market. Transforming capability has the role of maintaining firms' "evolutionary fitness" (Teece, 2007) and sustaining their competitive advantages.

2.2 Dynamic Capabilities in Sustainability Innovation Implementation

The provision of "adequate resources" is a critical driver of firms' sustainability initiatives (Schrettle et al., 2014). Particular skills, procedures, or routines are part of firms' resources that drive the success of sustainability innovation implementation (Cezarino et al., 2019). These skills, procedures, or routines can be systematically studied through the framework of DCs. In the context of sustainability innovation, DCs are organisational capabilities that set firms to sense and seize the opportunities or respond to threats stemming from evolving stakeholders' expectations about organisational sustainability practices (Wu et al., 2013). This, in turn, facilitates firms to achieve the triple bottom line of economic, environmental, and social benefits.

Firstly, sensing capabilities involve routines or procedures of obtaining knowledge about changing demands of customers and other stakeholders, strategic movements of competitors, relationships with suppliers, and other technological opportunities as well as other elements in the environment (Teece, 2007). Extant literature has emphasised the importance of knowledge about sustainability innovation opportunities, techniques, technologies, and market demands in implementing innovation towards sustainability (Souto & Rodriguez, 2015; Oliveira Neto et al., 2017; Segarra-Oña et al., 2016). These capabilities direct firms' focus towards the traditionally unattended sustainability requirements of stakeholders and detect opportunities that potentially bring along competitive advantages and threats that might negatively impact firms' operations. Through a comprehensive content analysis of CSR reports of leading UK companies, Wu et al. (2013) listed out some "microfoundations" (i.e. practices) belonging to sustainability sensing capabilities. These practices include the establishment of formal and informal communication channels with external stakeholders, participation in active dialogue with stakeholders regarding sustainability issues through meetings/conferences/newsletters, publishing strategic sustainability plans and receiving feedback from stakeholders and constant updates of information about new environmental requirements.

Secondly, seizing capabilities are seminal in achieving successful sustainability outcomes as there are possibilities that firms might succeed at sensing the opportunities or threats but fail to respond to them due to various reasons such as lack of commitment, risk aversion, or financial constraints for sustainability initiatives (Ellström et al., 2022). According to Teece (2007), seizing capabilities relate to routines and procedures to adjust and mobilise organisational resources to respond to sensed opportunities or threats. In particular, implementation of new products, services, or procedures is necessary. As engagement with sustainability innovation investment can be financially and managerially expensive (Jawahar & McLaughlin, 2001), in many cases, firms are unable to figure out how to achieve environmental and social outcomes as well as economic outcomes at the same time (McWilliams & Siegel, 2001). This situation might impede the implication of sustainability innovation. Therefore, to develop towards sustainability, it is important for firms to find the solutions to achieve all three goals simultaneously by effectively conducting interrelated activities that create, obtain, and combine a wide range of resources and competencies (Teece, 2007). From their analysis, Wu et al. (2013) emphasised the essence of defining a solid sustainability development strategy and governance structure as well as inter-organisational and intra-organisational knowledge-sharing routines. This is because, in comparison with normal market-driven innovation, sustainability-related innovation usually involves a wider set of stakeholders with conflicting interests (Hall & Vredenburg, 2004). Therefore, to seize the opportunities for sustainability innovation, firms must develop their cooperation with a wide network of partners (De Marchi, 2012). Some examples of "microfoundations" belonging to seizing cluster of DCs are developing clear strategic plans to execute the new sustainability initiatives, encouraging employees to share good practices and ideas about sustainability, maintaining teams to manage collaboration with external stakeholder groups in sustainabilities projects, and intra-organisational knowledge sharing about successful sustainability innovation practices (Wu et al., 2013).

Finally, transforming capabilities are defined as routines to facilitate renewal and reorchestration of resources and competencies to maintain firms' competitive advantage as firms continue to grow in a changing environment (Helfat & Peteraf, 2003; Teece, 2007). This cluster of DCs helps firms to avoid the "capability trap" (Teece et al., 1997; Zollo & Winter, 2002), where firms with superior competitive advantages have the tendency to persist with existing capabilities without acknowledging that their capabilities have been subverted under fast-changing market conditions (Leonard-Barton, 1992). For instance, firms should establish cross-functional teams to support sustainability innovation (Dangelico, 2016), perform regular measurement of efficiency levels of environmental activities, conduct continuous training for employees and suppliers about sustainability, and implement standardised environmental management systems (Wu et al., 2013).

2.3 *Critics of Dynamic Capabilities in Sustainability Innovations Studies*

While it has been recognised by scholars that DCs theory provides a systematic framework to understand how firms can navigate sustainability innovation implementation (Darmani et al., 2017; Amui et al., 2017), DCs theory is criticised for ignoring the context that DCs are conditioned and bounded (Barreto, 2010; Arend & Bromiley, 2009). This position downplays the role of institutions and other environmental forces, such as industry, culture, and socioeconomic factors, in defining firm behaviour (Gölgeci et al., 2017). Therefore, it might be insufficient to employ DCs theory as the sole framework to study sustainability innovation implementation. Having the appropriate capabilities does not ensure the success of sustainability innovation implementation as other factors might also impact this process. These factors can be internal ones that are inherent within the organisation or external ones to which firms are inevitably exposed to. For instance, in the process of developing a strategic plan for sustainability innovation, opinions from indirect stakeholders such as Non-Governmental Organisations (NGOs) are considered more and more important, besides those from direct stakeholders such as customers, shareholders, or suppliers (Wu et al., 2013). This is unfamiliar to firms (Hart & Dowell, 2011). Therefore, it is essential that managers to make deliberate decisions about establishing routines for searching and updating opinions from these stakeholders in the innovation process (Wu et al., 2013). However, DCs theory has not paid enough attention to the role of managers in the changing process (Augier & Teece, 2008). Furthermore, many external factors can affect the success of sustainability innovation implementation; these might include regulations and legislations, social norms and values, or implicit standards of the industry. For example, as governments around the world are trying to foster sustainability practices, favourable regulations might be put in place such as tax exemptions for firms which carry out sustainability innovations. This action can enable firms to release some financial burden while executing sustainability innovation projects and help them to achieve economic benefits along with environmental and social outcomes. As a result, those sustainability projects can be beneficial for all stakeholders and be highly likely to become successful.

Moreover, institutional factors can also shape the formation during the implementation of sustainability innovation. DCs have been criticised for being studied out of their relations with institutions that underpin and form their origins, nature as well as their mechanisms and influence their functions (Dunning & Lundan, 2008; Lessard et al., 2016). This ignorance might lead to inconsistent understanding of DCs (Barreto, 2010). Because the formation of DCs is greatly dependent on learning mechanisms (Zollo & Winter, 2002), the embedded institutional logics of organisations and internal actors joining the process can influence the formations and functions of DCs. According to Barreto (2010), four dimensions contribute to the formation of DCs; they are the propensity to change the resource base, to sense opportunities and threats, to make timely decisions, and to make market-oriented decisions. All of these involve cognitive processes of managers and other organisational actors, which are functioned by institutional logics of individuals and organisations.

In addition, institutional factors can influence how DCs function during sustainability innovation implementation. For instance, different logics can induce various interpretations among firms towards opportunities and threats sensed in the sensing phase. Towards the same external factor, one firm can perceive it as an opportunity while others consider it as a threat and execute different reactions later. This heterogeneity creates disequilibrium among firms (Teece, 2014) and helps some firms gain competitive advantages. Different reactions resulting from the interactions between institutions and DCs can affect the strategies pursued in the seizing phase. There are possibilities that firms might succeed at sensing the opportunities or threats but fail to respond to them due to several reasons, for example, lack of commitment, risk aversion, and financial constraints (Ellström et al., 2022). The first two of these barriers could be traced to underpinned institutional logics of actors or conflicting logics among those actors. For example, conflicting institutional logics between managers and staff can impinge on the possibility of a successful innovation implementation in the seizing phase when managerialism logics of managers promote efficiency, uniformity, and transparency in the new model while professionalism logics of professionals prefer autonomy, privacy, and personal reputation. This logic of professionals might act as a reversing force to hold back and might modify the actual outcome of the new business model compared with the original plan of managers.

Based on these discussions, this chapter is dedicated to explicating a framework that holistically investigates the sustainability implementation process in the current fast-changing business environment. We contend that DCs, institutional factors, and institutional logics can all impact Digital Transformation (DT); and institutional factors and institutional logics can ubiquitously impact DCs and how DCs function during sustainability innovation implementation process as follows:

(1) DCs, institutional forces and underlying institutional logics can influence sustainability innovation implementation and its success possibility.
(2) Institutional forces and institutional logics influence the formation of DCs (i.e. sensing, seizing, and transforming capabilities).
(3) Institutional forces and institutional logics can affect how DCs function during sustainability innovation implementation and its success possibility.

3 Neo-Institutional Theory in Sustainability Innovation Implementation Research

3.1 Brief Introduction About Neo-Institutional Theory

The origin of institutional theory is from the studies of Selznick (1948, 1949, and 1957) about institutions and organisational analysis (DiMaggio & Powell, 1983), which are classified as "old institutional" theory. These studies investigated organisations in the context of institutional environments and how organisations function in society rules, contracts, and authority. Later, Meyer and Rowan (1977) and Zucker (1977) commenced a neo-institutional theory with the focus on culture and cognition. The centre of the research from Meyer and Rowan (1977) was firms' rationality when they try to comply with external requirements for obtaining legitimacy. The work of Zucker (1977) concentrated more on the micro perspective to emphasise the taken-for-granted rules in organisations and the persistence of culture. Extending from these seminal works, DiMaggio and Powell (1983) moved the focus to the organisational field level to explicate the sources of isomorphism, which included coercive, mimetic, and normative forces. Based on this, Scott (1995) discussed about three pillars, which were regulative, normative, and cognitive pillars. While both DiMaggio and Powell (1983) and Scott (1995) emphasise the role of legitimacy in driving organisational action, their approaches are different. Scott (1995) seeded the three pillars as separable (i.e. the legitimacy of organisations came from either of the three pillars), while DiMaggio and Powell (1983) persuaded that the three isomorphism sources above collectively drive organisational actions. In this chapter, we adopt the view from DiMaggio and Powell (1983) as this approach focuses on the organisational field level which directly affects the environment of organisations. In addition, we contend that the three institutional forces collectively influence changes in organisation, as discussed in the work of DiMaggio and Powell (1983).

In the research of DiMaggio and Powell (1983), institutions were defined as taken-for-granted, culturally embedded "rules of the game" that govern and facilitate social and economic activity. These institutions enforce the order for social and economic life (Metcalfe, 2001) and they are enacted through coercive, normative, and mimetic mechanisms (DiMaggio & Powell, 1983). The focus of neo-institutional theory is to explain why organisations, as rational actors, adopt similar practices and try to change the organisations for survival. The underlying reason for this phenomenon is organisations' attempt to secure legitimacy which is considered the primary source of survival and success (DiMaggio & Powell, 1983).

Firstly, coercive force is from both formal and informal pressures imposed by regulatory entities or society (DiMaggio & Powell, 1983). Specifically, these forces can include statutory regulations like environmental protection standards, specific requirements on sustainability reports, etc. They can also be implicitly in the form of social norms and values accepted by the society in which firms operate. Coercive force can be clearly depicted through the study of Swidler (1979) which described the need to have the official role of "principal" in her free school where there is no need for administrative role. The reason is that her school cannot interact with other hierarchical organisations in the community without an "official principal".

Regarding mimetic force, this is the force that direct organisations' actions when they are in uncertainty. When the goals of organisations are ambiguous, or they have to operate in a disruptively changing environment, organisations feel the urge to mimic the moves of other organisations, which are seen as legitimate and successful. Organisations expect that if they model what perceived legitimate and successful organisations do, they can maximise their survival and success. Common channels for distributing ideas or practices among organisations are consultancy firms and industry associations (Bennich, 2024).

Finally, normative force is the force that results from professionalisation. The mechanism by which normative force functions is that during the recruitment process, personnel move from one organisation to another and bring with them professional norms, practices, and expectations which are adopted through universities or normative rules and professional behaviours diffused through professional associations. This process makes these norms and beliefs, expectations, behaviours and practices expand among different organisations within any organisational field.

3.2 Institutional Forces and Sustainability Innovation Implementation

Although isomorphism and homogeneity are still considered seminal to the study of institutional theory, later research has expanded the attention of institutional theory to institutional change (Bennich, 2024; Benner & Ranganathan, 2012). As the implementation of sustainability innovation can bring about transformative changes to the business models or processes of organisations, this process triggers a significant institutional change. Therefore, it is suggested that institutional theory can be useful in investigating the process of sustainability innovation implementation (Mohammad Ebrahimi & Koh, 2021; Bansal, 2005).

We suggest that future research should investigate how the three institutional forces developed by DiMaggio and Powell (1983) impact organisations' motivation to seek and develop sustainability innovation and to what extent these forces push organisations to commit to their plan. Coercive pressure plays a crucial role in sustainability practices management (Kilbourne et al., 2002). For example, regulatory requirements on the threshold of carbon emissions or increasing society's awareness of climate change can urge firms to seek innovation to reduce their carbon emissions. In addition, normative force can strongly drive organisational awareness about sustainability (Ball & Craig, 2010) as this pressure spread sustainability innovation practices among different organisations. In addition, as the process involved sustainability innovation is complex and sometimes confusing (Cezarino et al., 2019), organisations might have the tendency to replicate the sustainability practices of perceived successful organisations. Therefore, mimetic pressure can be a useful lens to investigate firms' actions towards sustainability innovation management.

Although the institutional forces might inclusively impact organisational change, we contend that the intensity level of each force is different among different organisational fields. For example, Bennich (2024) studied a water company in Sweden to investigate the presence of these three institutional forces for the motivation to adopt the "digital water" of the company. They concluded that mimetic and normative processes are the dominant pressures in this case, while coercive force merely presents. The case for other fields might be very dissimilar. Furthermore, we suggest that even when organisations operate in the same organisational field but in different regions, the intensity level of each force can also vary following specific regulatory or socio-cultural aspects of the regions. Therefore, exploring how the institutional forces behave in the sustainability innovation context for various organisational fields and geographical regions can be a fruitful area for future research and be meaningful for practitioners.

3.3 Impacts of Institutional Forces on the Formation of Dynamic Capabilities

Institutional factors can affect the process of forming new routines which are the building blocks of DCs (Dunning & Lundan, 2010) because these factors underpin how organisational actors exploit resources, capabilities, and markets available (Dunning & Lundan, 2008; Dunning, 2008). The three institutional forces can be influential contributors to the way DCs are created: which DCs are developed, sources of information for DCs to function, and how these DCs are formed. Firms aim to create DCs that maximise their survival in certain institutional contexts under the impacts of these forces. In this context, institutional forces should not be restrictively understood only as force; these forces might also be in the form of persuasion or invitation to join in collusion (DiMaggio & Powell, 1983). For instance, during the sensing phase, firms should actively detect opportunities and threats rising as the result of coercive force (i.e. new statutory regulations, new field standards institutionalised within the field, etc.). For example, in the context of sustainability innovation, as recycling is gaining in popularity and legitimised within society, some cosmetic producers have sensed this opportunity and developed routines to constantly update their recycling programmes to respond to this trend. Examples of these recycling programmes are discounts for customers who return their packages or refill products with old bottles. These programmes not only help firms to comply with societal expectations, but also enable them to enhance customer loyalty and save costs. Moreover, through the mimetic and normative process, firms form their DCs by adopting "best-practice" processes (i.e. DCs) from firms that are considered legitimate in the field or unintentionally adopt certain DCs brought about by individuals from other firms when hiring them. In the current borderless economy, these three forces are affecting firms more than ever before. Therefore, the study of DCs through the lens of institutional forces provides interesting directions for scholarly exploration. In particular, future research can investigate which institutional forces are present in each of the sensing, seizing, and transforming phases in a particular setting, how these forces foster or hinder the formation and function of DCs, and how firms can utilise the power of these forces to support the formation and function of DCs.

3.4 Impacts of Institutional Forces on the Functions of Dynamic Capabilities During Sustainability Innovation Implementation

Institutional forces can adjust the level of importance for a particular DC in its context. For instance, food safety and sustainably produced food are attracting enormous concern from customers and are also under strict scrutiny by the government. This situation creates a strong coercive force for companies in the food industry to comply with food safety standards and accelerate R&D activities to improve their sustainability practices in production. This force drives managers' attention to DCs that serve these targets. In this case, DCs relate to critical supply chain management practices for sustainability in the food industry such as triple bottom line capability, standards and certifications management capability, and life cycle assessment capability (Beske et al., 2014) might be even more important than DCs for developing tasty products which are commonly perceived as important in food production. Therefore, a novel avenue for future research is to investigate which are the

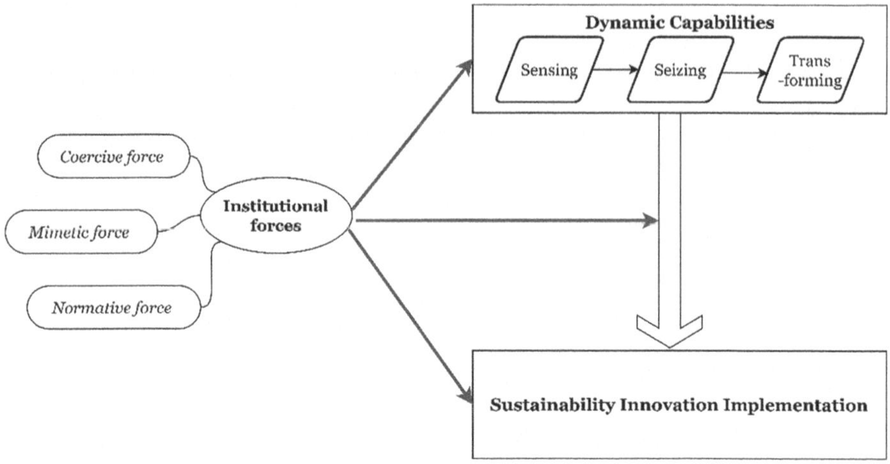

Fig. 2 Impacts of institutional forces on sustainability innovation implementation, formation of DCs and functions of DCs during sustainability innovation implementation

institutional forces that critically affect the survival and the ability to gain competitive advantages of firms in a particular context. From that, appropriate DCs might be recommended for those companies as it might be the case that the organisations select to develop DCs that are not the worthiest in their situation.

Figure 2 summarises the proposed impacts of institutional forces on sustainability innovation implementation, the formation of DCs and the functions of DCs during sustainability innovation implementation.

4 Institutional Logics and Sustainability Innovation Implementation

4.1 Brief Introduction About Institutional Logics Perspective

Deriving from the original institutional theory, the institutional logics perspective is a seminal perspective that concerns how organisational structure is shaped by cultural and cognitive factors (Thornton & Ocasio, 2008). The foundation of institutional logics perspective is set by several seminal works from Friedland and Alford (1991), Haveman and Rao (1997); Thornton and Ocasio (1999); Scott et al. (2000), and Thornton and Ocasio (2008).

Institutional logics are defined as "the set of material practices and symbolic systems including assumptions, values, and beliefs by which individuals and organisations provide meaning to their daily activity, organise time and space, and reproduce their lives and experiences". Institutional logics manifest at various levels from societal level, field level, organisational level, and individual level (Thornton &

Ocasio, 2008). Scholars have categorised seven institutional orders and their associated logics, which are family, community, religion, state, market, profession, and corporation (Friedland & Alford, 1991; Thornton, 2004; Thornton et al., 2012). The institutional logics at other levels are founded on the basis of these seven societal-level institutional logics. In the extant literature, institutional logics perspective has been employed to delineate institutional changes in institutional systems (Thornton et al., 2012; Thornton & Ocasio, 1999).

4.2 Institutional Logics and Sustainability Innovation Implementation

Extant research has been attempted to study the roles of different institutional logics and their conflicts on organisations' sustainability practices (e.g. Olesson et al. (2023); Corbett et al. (2018); Ozdamar Ertekin et al. (2020); Milosevic et al. (2023)). Institutional logics at different levels (i.e. societal, organisational field, organisational, and individual level) can shape sustainability innovation direction of organisations.

Firstly, institutional logics at the societal level and organisational field level can immensely affect the expected actions or behaviours of organisations regarding sustainability practices. These expectations will direct how organisations design their sustainability alternatives (Arena et al., 2018). In particular, the institutional logics at these macro levels have the power to shape the social obligations that firms must comply with and determine the expected characteristics of sustainability activities (Arena et al., 2018). For example, Marano and Kostova (2016) found that the national context affects institutional logics that create heterogeneity in the adoption of sustainability practices across different countries. In another research by Ozdamar Ertekin et al. (2020), it was stated that the dominance of commerce logic in fashion field has brought about many negative impacts on environment and society. In the context of increasing natural disasters, emergence of the new "logic of sustainability" is necessary. Together with the logic of art in the fashion field, the three logics exist simultaneously and the intersections among them both support and challenge the institutionalisation of a sustainable fashion system.

Secondly, the institutional logics at organisational and individual level will directly shape how sustainability innovation implementation evolves. The outcomes of the sustainability innovation implementation are the result of managerial strategic decisions and actions of lower-level actors following those decisions. Both managers' strategic decisions and how other organisational actors carry out the strategic plan are enabled and shaped by prevailing institutional logics. This is because the interests, identities, values, and assumptions of individuals together shape their choices and actions; and all these factors are embedded within prevailing institutional logics of individuals (Thornton & Ocasio, 2008). In specific, institutional logics of key decision-makers can direct their focus to a certain set of issues and solutions (Joseph & Wilson, 2018; Ocasio, 1997) and decision-makers' attention, in turn, affects how organisations respond to social and economic factors. Institutional logics provide individuals with a set of values which defines "the legitimacy, importance and relevance of issues and solutions" and enable individuals to be aware of their "interests and identities" (Thornton & Ocasio, 2008). For example, among different alternatives for solving a sustainability issue, managers who hold interests in technology solutions might be more likely to approach technological innovation solutions than directors who are technophobe.

Extant literature has focused on investigating institutional logics in sustainability practice from only single-level approach. However, we believe that sustainability attempts should not be understood through a single level of institutional logic. Otherwise, this process should be examined under the collective impacts of all aforementioned levels. Furthermore, since institutional logics is context-dependent, a novel research direction is to attempt to study sustainability innovation through contextualised institutional logics in different contexts (different geographical areas, periods, organisational fields, etc.).

4.3 Institutional Logics and the Formation of Dynamic Capabilities

In addition, we contend that institutional logics can also affect the formation and functions of DCs. For studying institutional logics in relationship with DCs, we suggest that the focus should be on the organisational and managerial individual institutional logics as these logics directly impact the formation and functions of such strategic-level capabilities as DCs. Firstly, as institutional logics of a particular organisation define what are considered valuable and right for that organisation, organisational logics identify the outcomes that DCs are built to achieve and determine if a particular DC is acceptable and can be maintained within that organisation. Secondly, as articulated in the previous section, prevailing institutional logics set the locus of decision-makers' attention to certain issues and solutions (Joseph & Wilson, 2018; Ocasio, 1997). That means in the sensing phase, managerial institutional logics would direct their focus towards certain opportunities and threats which are the basis for deciding what microfoundation DCs are needed. Subsequently, managers would tend to select certain solutions under their logics to address the opportunities and threats sensed in the first phase. These solutions require a specific set of microfoundations DCs to be established. This behaviour repeats in the transformation phase.

In the literature of business and management, institutional logics are studied at different levels from societal level to individual level and at various levels of complexity of interaction. By level of complexity of interaction, extant research has investigated single dominant institutional logics and multiple institutional logics (Wu et al., 2023) within organisations. We suggest researchers should investigate the formation and function of DCs under the influence of multiple logics as it is rare that only one institutional logic is dominant in our current dynamic environment. Future research should explore how organisational and managerial individual logics together direct the formation and functions of digital DCs and how to establish digital DCs that harmonise with organisational logics.

4.4 *Impacts of Institutional Logics on the Function of Dynamic Capabilities During Sustainability Innovation Implementation*

As DCs are routines and processes (Barreto, 2010) operationalised by organisational actors, individual institutional logics of managers and relevant staff would affect the process and the output of that process, which are the ingredients for sustainability innovation implementation. For example, in the sensing phase, institutional logics of people who join this phase act as the lens for them to scan the environment and determine what are the opportunities and threats for their organisations. Therefore, institutional logics will affect the output of the sensing process and set the foundation for the direction of sustainability innovation design. After a new opportunity is discovered, it must be responded to with a new product or service, or even a new business model (Teece, 2007). The selection of which product, service, or business model reflects managers' interpretation of the opportunities and their expectations about customers' demand and how the business delivers these values to customers and gets payment from them. Institutional logics of managers play a crucial role in this process as it can shape managers' thoughts and interpretations of things. In the transformation phase, organisations continue to reconfigure their resources to sustain their evolutionary fitness as the enterprises grow. Similar to the previous phase, this process is also affected by managers' institutional logics.

Furthermore, different institutional logics can affect the level of importance of certain DCs in the sustainability innovation implementation process. For example, DCs that serve to increase the professional skillfulness of staff might be more crucial in case managers hold the logic that focuses on professional capabilities. In contrast, if the managers have the logic that focuses on profitability, DCs that relate to seizing new customers or cost efficiency might be more predominant. The problem is whether those particular predominant DCs are truly meaningful towards the sustainability plan of the firms, or they just arise out of the subjective logic of the decision-makers. This is a meaningful question for future research to investigate and the answer to this would be highly valuable for organisations to build meaningful DCs for their sustainability innovation implementation.

Figure 3 summarises the influences of institutional logics on sustainability innovation implementation, the formation of DCs and the functions of DCs during sustainability innovation implementation.

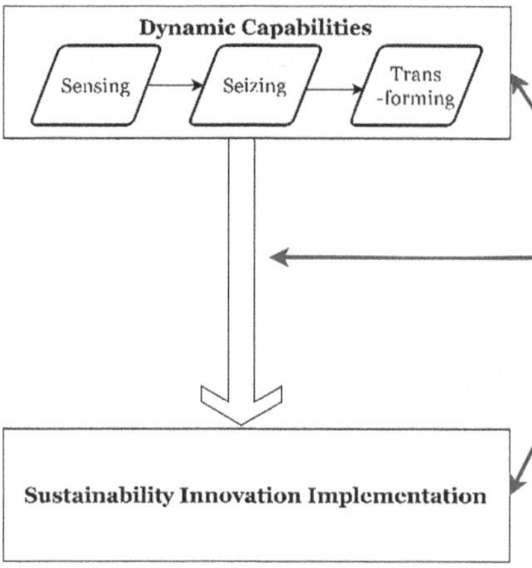

Fig. 3 Impacts of institutional logics on sustainability innovation implementation, formation of DCs and functions of DCs during sustainability innovation implementation

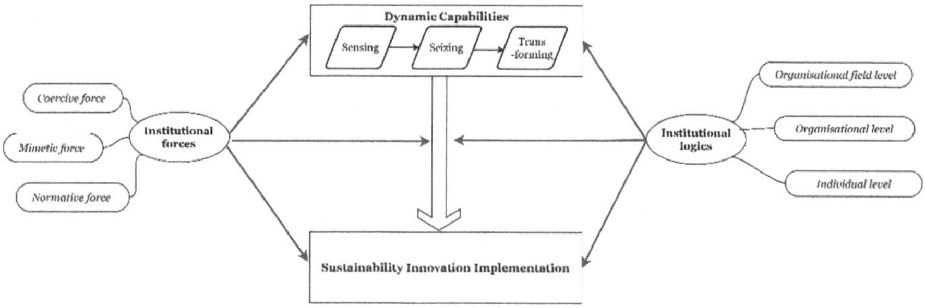

Fig. 4 Conceptual framework for dynamic capabilities, institutional forces, institutional logics, and sustainability innovation implementation

5 A Framework for Institutional Forces, Institutional Logics, and Dynamic Capabilities and Sustainability Innovation Implementation

As explicated in the previous sections, dynamic capabilities theory, neo-institutional theory, and institutional logics perspective can provide foundational basis for organisations to understand, analyse, and enhance the outcomes of their sustainability innovation implementation. These theories and perspectives complement each other to draw a comprehensive picture of factors that affect sustainability innovation design and implementation as well as how organisations can navigate the changes associated with sustainability innovations. Fig. 4 depicts the conceptual framework about how institutional forces, institutional logics, and dynamic capabilities interact and impact sustainability innovation implementation.

6 Conclusion

Sustainability innovations are crucial for the sustainability development of organisations. However, extant literature about models for sustainability innovations still lacks solid theoretical foundation (Dangelico et al., 2017). This chapter proposes a comprehensive framework for firms to understand, analyse, and enhance their sustainability innovations implementation to achieve social, environmental benefits while still ensuring economic benefits.

In our proposed framework, we employ DCs theory, neo-institutional theory, and institutional logics perspective to develop a conceptual framework to comprehensively analyse and implement sustainability innovations. As sustainability innovations usually involve revolutionary changes in product designs, organisational practices or processes (Adams et al., 2016), dynamic capabilities theory has gained growing attention from scholars as this theory is dedicated to guiding organisations to integrate, build, and reconfigure their resources to gain competitive advantages in fast-changing environments. However, DCs theory is often criticised about its ignorance of contextual factors that impact the DCs themselves and organisations' operations (Barreto, 2010). To address these critiques, in our conceptual framework, we integrate neo-institutional theory (DiMaggio & Powell, 1983) and institutional logics perspective (Friedland & Alford, 1991; Thornton & Ocasio, 2008; Thornton et al., 2012) to account for the institutional factors that shape organisational actions. On the other hand, DCs theory also accounts for acknowledgement of firms' abilities to innovate and gain competitive advantages which is overlooked in institutional theory (Greenwood et al., 2017; Deephouse et al., 2017; Suddaby & Viale, 2011).

In particular, dynamic capabilities enable firms to identify and respond to the opportunities and threats arising from changing stakeholders' demand in the context of increasing awareness about sustainability. In addition, we propose that institutional forces (i.e. coercive, mimetic, and normative forces), as well as institutional logics at different levels (i.e. societal, organisational field, organisational, and individual levels), affect sustainability innovation implementation, the formation of DCs and the functions of DCs during sustainability innovation implementation. These notions can open new research opportunities for a more holistic study of sustainability innovations. Although there is an increasing number of sustainability innovation research employing institutional theory, this chapter is one of the first to suggest a combination of DCs theory and institutional theory for a framework containing the action part (DCs theory), the contextual part (institutional theory), and the possible effects brought by the interaction between institutional factors and DCs during the process. The authors expect that this view can address the critical comments raised by researchers regarding the use of DCs theory and institutional theory. In addition, we believe that this conceptual framework can facilitate a more comprehensive understanding of the sustainability innovation implementation process in different contexts (different organisational fields, geographic areas, etc.). Besides the theoretical contributions, this chapter also informs practitioners about critical factors that might affect the success of sustainability innovation, especially the factors that they might be unaware of like institutional logics. Future research can either use the proposed framework as a whole to empirically study sustainability implementation cases or break down the framework into various research directions.

References

Adams, R., Jeanrenaud, S., Bessant, J., Denyer, D., & Overy, P. (2016). Sustainability-oriented innovation: A systematic review. *International Journal of Management Reviews, 18*(2), 180–205.

Amui, L. B. L., Jabbour, C. J. C., de Sousa Jabbour, A. B. L., & Kannan, D. (2017). Sustainability as a dynamic organizational capability: A systematic review and a future agenda toward a sustainable transition. *Journal of Cleaner Production, 142*, 308–322.

Arena, M., Azzone, G., & Mapelli, F. (2018). What drives the evolution of corporate social responsibility strategies? An institutional logics perspective. *Journal of Cleaner Production, 171*, 345–355.

Arend, R. J., & Bromiley, P. (2009). Assessing the dynamic capabilities view: Spare change, everyone? *Strategic Organization, 7*(1), 75–90.

Augier, M., & Teece, D. J. (2008). Strategy as evolution with design: The foundations of dynamic capabilities and the role of managers in the economic system. *Organization Studies, 29*(8–9), 1187–1208.

Ball, A., & Craig, R. (2010). Using neo-institutionalism to advance social and environmental accounting. *Critical Perspectives on Accounting, 21*(4), 283–293.

Bansal, P. (2005). Evolving sustainably: A longitudinal study of corporate sustainable development. *Strategic Management Journal, 26*(3), 197–218.

Bansal, P., & Roth, K. (2000). Why companies go green: A model of ecological responsiveness. *Academy of Management Journal, 43*(4), 717–736.

Barreto, I. (2010). Dynamic capabilities: A review of past research and an agenda for the future. *Journal of Management, 36*(1), 256–280.

Benner, M. J., & Ranganathan, R. (2012). Offsetting illegitimacy? How pressure from securities analysts influence incumbents in the face of new technologies. *Academy of Management Journal, 55*(1), 213.

Bennich, A. (2024). The digital imperative: Institutional pressures to digitalise. *Technology in Society, 76*, 102436.

Beske, P., Land, A., & Seuring, S. (2014). Sustainable supply chain management practices and dynamic capabilities in the food industry: A critical analysis of the literature. *International Journal of Production Economics, 152*, 131–143.

Cezarino, L. O., Alves, M. F. R., Caldana, A. C. F., & Liboni, L. B. (2019). Dynamic capabilities for sustainability: Revealing the systemic key factors. *Systemic Practice and Action Research, 32*(1), 93–112.

Corbett, J., Webster, J., & Jenkin, T. A. (2018). Unmasking corporate sustainability at the project level: Exploring the influence of institutional logics and individual agency. *Journal of Business Ethics, 147*(2), 261–286.

Dangelico, R. M. (2016). Green product innovation: Where we are and where we are going. *Business Strategy and the Environment, 25*(8), 560–576.

Dangelico, R. M., Pujari, D., & Pontrandolfo, P. (2017). Green product innovation in manufacturing firms: A sustainability-oriented dynamic capability perspective. *Business Strategy and the Environment, 26*(4), 490–506.

Darmani, A., Niesten, E. M. M. I., & Hekkert, M. P. (2017). Characteristics of investors in onshore wind power in Sweden. *Environmental Innovation and Societal Transitions, 24*, 67–82.

De Marchi, V. (2012). Environmental innovation and R&D cooperation: Empirical evidence from Spanish manufacturing firms. *Research Policy, 41*(3), 614–623.

Deephouse, D. L., Bundy, J., Tost, L. P., & Suchman, M. C. (2017). Organizational legitimacy: Six key questions. In *The SAGE handbook of organizational institutionalism* (p. 27). SAGE Publications Ltd.

DiMaggio, P. J., & Powell, W. W. (1983). The iron cage revisited: Institutional isomorphism and collective rationality in organizational fields. *American Sociological Review, 48*(2), 147–160.

Dunning, J. H. (2008). *Multinational enterprises and the global economy* (2nd ed., S. M. Lundan, Ed.) Edward Elgar.

Dunning, J. H., & Lundan, S. M. (2008). Institutions and the OLI paradigm of the multinational enterprise. *Asia Pacific Journal of Management, 25*(4), 573–593.

Dunning, J. H., & Lundan, S. M. (2010). The institutional origins of dynamic capabilities in multinational enterprises. *Industrial and Corporate Change, 19*(4), 1225–1246.

Eisenhardt, K. M., & Martin, J. A. (2000). Dynamic capabilities: What are they? *Strategic Management Journal, 21*(10–11), 1105–1121.

Ellström, D., Holtström, J., Berg, E., & Josefsson, C. (2022). Dynamic capabilities for digital transformation. *Journal of Strategy and Management, 15*(2), 272–286.

Friedland, R., & Alford, R. R. (1991). Bringing society back in: Symbols, practices, and institutional contradictions. In W. W. Powell & P. J. DiMaggio (Eds.), *The new institutionalism in organizational analysis* (pp. 232–266).

Gölgeci, I., Larimo, J., & Arslan, A. (2017). Institutions and dynamic capabilities: Theoretical insights and research agenda for strategic entrepreneurship. *Scandinavian Journal of Management*, 33(4), 243–252.

Greenwood, R., Oliver, C., Lawrence, T. B., & Meyer, R. E. (2017). *The SAGE handbook of organizational institutionalism* (2nd ed.). SAGE Publications.

Hall, J., & Vredenburg, H. (2004). The challenges of innovating for sustainable development. *MIT Sloan Management Review, 45*(1), 61–68.

Harsanto, B., Kumar, N., & Michaelides, R. (2024). Sustainability-oriented innovation in manufacturing firms: Implementation and evaluation framework. *Business Strategy and the Environment, 33*(6), 5086–5108.

Hart, S. L., & Dowell, G. (2011). A natural-resource-based view of the firm: Fifteen years after. *Journal of Management, 37*(5), 1464–1479.

Haveman, H. A., & Rao, H. (1997). Structuring a theory of moral sentiments: Institutional and organizational coevolution in the early thrift industry. *American Journal of Sociology, 102*(6), 1606–1651.

Helfat, C. E., & Peteraf, M. A. (2003). The dynamic resource-based view: Capability lifecycles. *Strategic Management Journal, 24*(10 SPEC ISSUE), 997–1010.

Jawahar, I. M., & McLaughlin, G. L. (2001). Toward a descriptive stakeholder theory: An organizational life cycle approach. *The Academy of Management Review, 26*(3), 397–414.

Joseph, J., & Wilson, A. J. (2018). The growth of the firm: An attention-based view. *Strategic Management Journal, 39*(6), 1779–1800.

Kilbourne, W. E., Beckmann, S. C., & Thelen, E. (2002). The role of the dominant social paradigm in environmental attitudes: A multinational examination. *Journal of Business Research, 55*(3), 193–204.

Leonard-Barton, D. (1992). Core capabilities and core rigidities: A paradox in managing new product development. *Strategic Management Journal, 13*(S1), 111–125.

Lessard, D., Teece, D. J., & Leih, S. (2016). Introduction to special topic forum on developing the dynamic capabilities of global companies across levels and locations. *Global Strategy Journal, 6*(3), 165–167.

Marano, V., & Kostova, T. (2016). Unpacking the institutional complexity in adoption of CSR practices in multinational enterprises. *Journal of Management Studies, 53*(1), 28–54.

Martinuzzi, A., & Krumay, B. (2013). The good, the bad, and the successful – How corporate social responsibility leads to competitive advantage and organizational transformation. *Journal of Change Management, 13*(4), 424–443.

McWilliams, A., & Siegel, D. (2001). Corporate social responsibility: A theory of the firm perspective. *The Academy of Management Review, 26*(1), 117–127.

Metcalfe, J. S. (2001). Institutions and progress. *Industrial and Corporate Change, 10*(3), 561–586.

Meyer, J. W., & Rowan, B. (1977). Institutionalized organizations: Formal structure as myth and ceremony. *American Journal of Sociology, 83*(2), 340–363.

Milosevic, I., Bass, A. E., & Schulte, B. (2023). The interplay of conflicting and complementing institutional logics in sustainability practices. *Management International Review, 63*(3), 469–506.

Mohammad Ebrahimi, S., & Koh, L. (2021). Manufacturing sustainability: Institutional theory and life cycle thinking. *Journal of Cleaner Production, 298*, 126787.

Mousavi, S., Bossink, B., & van Vliet, M. (2018). Dynamic capabilities and organizational routines for managing innovation towards sustainability. *Journal of Cleaner Production, 203*, 224–239.

Nguyen, H. T. T., Ullah, S., Le, H. T. M., & Hameed, A. (2023). Sustainability targets in executive remuneration contracts and corporate sustainability performance in the United Kingdom and European Union. *Environment Systems and Decisions, 43*(3), 393–415.

Obermann, J., & Velte, P. (2018). Determinants and consequences of executive compensation-related shareholder activism and say-on-pay votes: A literature review and research agenda. *Journal of Accounting Literature, 40*(1), 116–151.

Ocasio, W. (1997). Towards an attention based view of the firm. *Strategic Management Journal, 18*(S1), 187–206.

Olesson, E., Nenonen, S., & Newth, J. (2023). Enablers and barriers: The conflicting role of institutional logics in business model change for sustainability. *Organization and Environment, 36*(2), 228–252.

Oliveira Neto, G. C., Leite, R. R., Shibao, F. Y., & Lucato, W. C. (2017). Framework to overcome barriers in the implementation of cleaner production in small and medium-sized enterprises: Multiple case studies in Brazil. *Journal of Cleaner Production, 142*, 50–62.

Ozdamar Ertekin, Z., Atik, D., & Murray, J. B. (2020). The logic of sustainability: Institutional transformation towards a new culture of fashion. *Journal of Marketing Management, 36*(15–16), 1447–1480.

Sargut, G., & McGrath, R. G. (2011). Learning to live with complexity. *Harvard Business Review, 89*(9), 68–136.

Schilke, O., Hu, S., & Helfat, C. E. (2018). Quo Vadis, dynamic capabilities? A content-analytic review of the current state of knowledge and recommendations for future research. *Academy of Management Annals, 12*(1), 390–439.

Schrettle, S., Hinz, A., Scherrer-Rathje, M., & Friedli, T. (2014). Turning sustainability into action: Explaining firms' sustainability efforts and their impact on firm performance. *International Journal of Production Economics, 147*(PART A), 73–84.

Scott, W. R. (1995). *Institutions and organizations*. Sage.

Scott, W. R., Martin, R., Peter, M., & Carole, C. (2000). *Institutional change and healthcare organizations: From professional dominance to managed care*. University of Chicago Press.

Segarra-Oña, M., Peiró-Signes, Á., & Mondéjar-Jiménez, J. (2016). Twisting the twist: How manufacturing & knowledge-intensive firms excel over manufacturing & operational and all service sectors in their eco-innovative orientation. *Journal of Cleaner Production, 138*, 19–27.

Souto, J. E., & Rodriguez, A. (2015). The problems of environmentally involved firms: Innovation obstacles and essential issues in the achievement of environmental innovation. *Journal of Cleaner Production, 101*, 49–58.

Suddaby, R., & Viale, T. (2011). Professionals and field-level change: Institutional work and the professional project. *Current Sociology, 59*(4), 423–442.

Swidler, A. (1979). *Organization without authority*. Harvard University Press.

Teece, D. J. (2007). Explicating dynamic capabilities: The nature and microfoundations of (sustainable) enterprise performance. *Strategic Management Journal, 28*(13), 1319–1350.
Teece, D. J. (2014). The foundations of enterprise performance: Dynamic and ordinary capabilities in an (economic) theory of firms. *Academy of Management Perspectives, 28*, 328–352.
Teece, D. J., Pisano, G., & Shuen, A. (1997). Dynamic capabilities and strategic management. *Strategic Management Journal, 18*(7), 509–533.
Thornton, P. H. (2004). *Markets from culture: Institutional logics and organizational decisions in higher education publishing*. Stanford Business Books.
Thornton & Ocasio. (2008). *Institutional logics*. SAGE.
Thornton, P. H., & Ocasio, W. (1999). Institutional Logics and the Historical Contingency of Power in Organizations: Executive Succession in the Higher Education Publishing Industry, 1958–1990. *The American Journal of Sociology, 105*(3), 801–843.
Thornton, P. H., Ocasio, W., & Lounsbury, M. (2012). *The institutional logics perspective*. Oxford University Press.
Vilanova, M., Lonzano, J. M., & Arenas, D. (2009). Exploring the nature of the relationship between CSR and competitiveness. *Journal of Business Ethics, 87*(Suppl 1), 57–69.
Winter, S. G. (2003). Understanding dynamic capabilities. *Strategic Management Journal, 24*(10 SPEC ISSUE), 991–995.
Wu, Q., He, Q., & Duan, Y. (2013). Explicating dynamic capabilities for corporate sustainability. *EuroMed Journal of Business, 8*(3), 255–272.
Wu, X., Tan, X., & Wang, X. (2023). The institutional logics perspective in management and organizational studies. *Journal of Business Research, 167*, 114183.
Zollo, M., & Winter, S. G. (2002). Deliberate learning and the evolution of dynamic capabilities. *Organization Science, 13*(3), 339–351.
Zucker, L. G. (1977). The role of institutionalization in cultural persistence. *American Sociological Review, 42*(5), 726.

Innovation and Competitive Advantage: Leading the Sustainable Revolution in Agricultural Entrepreneurship

Justina Teronpi, Juthika Konwar, and Nilanjana Chakraborty

Abstract Sustainable agricultural entrepreneurship is critical for food security and biodiversity, as global food demand is expected to reach 70% by 2050. Sustainability is critical for addressing problems such as food insecurity, natural disasters, diseases, and unemployment. The term "sustainable" refers to a person's ability to handle stress and recover from shocks, while not undermining the natural resource base. The agricultural sector, which accounts for 4% of global GDP, is facing challenges with extreme clumsiness and risks, climate change, and environmental resource degradation. Sustainable agricultural practices possess the ability to reduce the consumption of natural resources while improving the quality of biodiversity. Agricultural entrepreneurship is on the rise, with a focus on farm and non-farm activities to diversify income streams and address challenges. Innovative practices such as vertical farming, hydroponics, greenhouse farming, and organic farming are being used to ensure food security while preserving the natural resource base. Farmers face challenges such as technology costs, technical know-how, farm mechanisation, indifference, a lack of support, monitoring, and awareness. Despite these challenges, sustainable practices have the potential to bridge the economic and biodiversity gaps, ensuring food security and sustainable agriculture.

1 Introduction

India is an agrarian economy where about 55% of the population of India is dependent on agriculture and its allied activities for their daily livelihood. The agricultural sector also remains an important base for the supply of raw materials which is vital for business and especially for many developing countries, it is the main source of livelihood in developing countries especially in India. However, this sector is perceived to be associated with adversities, clumsiness, and extreme risk mostly because of the lack of information, resources, skills, and infrastructure without

J. Teronpi · J. Konwar · N. Chakraborty (✉)
Assam University, Silchar, India

A. Sharma et al. (eds.), *The Synergistic Effect of Sustainable Business Practices on Corporate Performance*, https://doi.org/10.1007/978-981-96-3375-3_3

which it is not possible to develop new ventures. The emergence of free market globally has led to the emergence of a new enterprise known as "Agripreneurship" which is defined as a "generally, sustainable, community-oriented, directly-marketed agriculture". An agripreneur is an individual who identifies viable opportunities in the agricultural and allied sectors in a business environment. In India, agribusiness formally came into being after the commencement of Agribusiness and Agriclinics Scheme (ACABC) in 2002 by the Ministry of Agriculture and Farmers' Welfare (MoA and FW), Government of India, to provide agricultural services and self-employment opportunities to farmers (Afroz et al., 2022). These "agripreneurs" have shown that given a chance they can transform Indian agriculture to meet with emerging needs (Parambil, 2016). In the aspect of mounting joblessness and dearth in rural areas, there is the need for entrepreneurship in agriculture. Agripreneurs are not just farmers but they are thinkers, risk takers, and business-minded individuals who are always on the look-out for opportunities. Hence, agricultural entrepreneurship is not just expanding farm business but including non-agricultural activities along with agriculture (Kumari and Kumar, n.d.). With several concerns, like food shortages, natural catastrophes, illnesses, unemployment, etc., becoming more intense every year, sustainability is unquestionably a must. Sustainability can only be achieved when the resources are optimised and the person can bounce back from shocks and strains both now and in the future. Additionally, it provides people with the tools they need to maintain themselves and take care of their fundamental necessities (Kabir et al., 2012). A sustainable lifestyle includes five indicators: income, enhanced well-being, decreased vulnerability, greater food security, and more sustainable resource use. People can satisfy their requirements if they have a sufficient income. Due to globalisation and the deregulation of the agricultural industry, it is crucial for any country's growth to maintain the sustainability of farmers today. Agri-entrepreneurship is currently promoted as the solution to empowering people in the rural regions by enhancing the livelihood of resource-deplete farmers. Studying the innovation led by sustainability in agriculture in this setting is crucial in determining how much it has aided the farmers and entrepreneurs in overcoming obstacles and rising beyond their precarious circumstances.

2 Materials and Methods

This study is aimed at highlighting innovation in agricultural entrepreneurship and the advantage of implementing sustainable practices to mitigate challenges posed by years of degradation in biodiversity caused by human practices and ignorance. Through a comprehensive review of literature through journal articles, books, websites, and reports, we seek to bring attention the need for adopting sustainable practices in the following sections where we discussed agricultural entrepreneurship, sustainability and in agricultural entrepreneurship, certain changes brought by entrepreneurship and competitive advantage of sustainable practices.

3 Discussions

3.1 Agricultural Entrepreneurship

The concept of "agripreneurship" dates back to the mid-eighteenth century; where Richard Cantillon and Jean Baptiste coined the term "entrepreneur" for a farmer who undertakes to pay the owner, for his farm or land a fixed amount of money with no guarantee of obtaining profits whatsoever and the latter associated the term to a farmer who is not only a risk bearer but coordinates all the activities of production in a farm (Dias et al., 2019). Ironically, there is a dearth in the mainstream plethora of research surrounding entrepreneurship in agriculture. The origins suggest that the main founding fathers of entrepreneurship postulated the theories when agriculture was the main industry and examples were cited frequently from the erstwhile existing agricultural practices in those eras. Agripreneurship relates to establishing innovative enterprises for the sake of making profits while fully estimating the risks associated with it and in this context, agripreneurs are those entrepreneurial individuals in the agriculture and the allied sectors who create and run their own ventures (Yoganandan et al., 2022). Agripreneurs take up value addition to agricultural products as their main activity and the starting point of their business activities. They also change the way of conducting work and gets involved in activities that will increase their agricultural output (Vidani et al., 2017). Agripreneurial activities are mainly classified into agricultural input, production, processing, creating and maintaining agribusiness infrastructure, and trading management and marketing (Brahma & Tripathi, 2020; Kadirvel et al., 2020; Kumari & Kumar, n.d.; Uplaonkar & Biradar, 2015). Thus, agripreneurs are new-age farmers that bring together farming and agriculture with business. But all agripreneurs are not farmers, some agripreneurs add value to the products that the farmers have grown. Farmers, on the other hand, produce keeping in mind their consumption needs and the MSP (minimum support price). Agripreneurship is thus, a union with entrepreneurship that turns farms into agribusiness (Lalhmingthanga et al. 2020). Therefore, an agripreneur works within the boundary of a food system, not necessarily in a rural location that sees an opportunity within a market to directly produce food stuffs, such as vegetables, meat, fish, and grains utilising innovative and sustainable production methods (Carr & Roulin, 2016; Magagula et al., 2020). Thus, like all entrepreneurs, agripreneurs also take risks, are visionaries, and are ready to start new ventures (Vidani et al., 2017) (Table 1).

3.2 Sustainability in Agricultural Entrepreneurship

Sustainability is essential due to rising issues such as food insecurity, natural disasters, diseases, and unemployment. The term "Sustainable" denotes a person's capacity to provide for themselves in the long run. De Haan (2000) defines it as a person's

Table 1 Some definitions of Agripreneurship/Agricultural entrepreneurship by different authors

Author	Definition
Dollinger (2003)	Entrepreneurship in agriculture as the creation of organisations for the purpose of growth or gain under conditions of risk and uncertainty in agriculture.
Dabson and Markley (2010)	An agripreneur is an individual whose main business activity is agriculture or allied sectors.
Carr and Roulin (2016)	An agripreneur works within the boundary of a food system, not necessarily in a rural location that sees an opportunity within a market to directly produce food stuffs, such as vegetables, meat, fish, and grains utilising innovative and sustainable production methods.
Dhingra et al. (2018); Parambil (2016)	Agripreneurship stemmed from the need for farmers to redefine agriculture and run their own enterprises.
Vidani et al. (2017)	An agripreneur adds values to agricultural products by starting his/her own business, changing the way of conducing work and getting involved in activities that will increase his or her agricultural output.
Chand (2019)	Agripreneurship is the entrepreneurial process taken up in agriculture and the allied sectors.
Lalhmingthanga et al. (2020)	Agripreneurship is the union of farming and entrepreneurship that transforms farms into agribusiness.
Brahma and Tripathi (2020)	"Agripreneurs" or "Agri-entrepreneurs" are those individuals whose "main business is agricultural-based or agriculture".
Falola et al. (2022)	Agripreneurship is the identification, development, and management of opportunities in agriculture in an optimal manner in order to make profit.
Agu (2020)	Entrepreneurship in agriculture is defined as agripreneurship.

ability to manage with and recuperate from stresses and shocks and sustain or boost their capabilities, assets, and activities both in the present and in the future. "Sustainability" as a word was not made popular till the 1980s, where there was an explosion to the usage of this term, mainly surging from the Brundtland Commission on Environment and Development in 1987 (Caradonna, 2014; De Haan, 2000; Zhang et al., 2019). Since the 80's, there have been several theories and frameworks postulated to understand the and define in order to meet the growing demands of sustainability in the environment where social scientists; namely Jeffrey D. Sachs have deemed human beings as "another form of natural disaster" who is "dominating over the natural environment". Some of the frameworks that have emerged include the Sustainable Livelihood Framework, the Millennium Development Goals, and, more recently, the Sustainable Development Goals all of which seem to have been developed with the idea of responding to the recent adversities faced around the world; mostly created by humans themselves. But the central theme remains the same: enable and equip individuals in the long run so as not to undermine the natural resource base. However, in order to achieve sustainability, one of the critical aspects is to have a sustainable food system. Therefore, sustainability in agricultural systems need to be promoted and developed in order to meet the sustainable goals. The agricultural industry is one of the largest sectors in the world which accounts for 4% of the global gross domestic product and employs more than 1 billion people (World Bank, 2024; World Wildlife Fund, n.d.). Subsistent agricultural practices, climate change, multiple shocks and stresses in the environment such as conflicts, disruptions in the weather conditions, market liberalisation, modernisation of agriculture, and decreased level of financing are impacting agricultural growth and development. All of these are having an effect on people's health, increasing food insecurity, and forcing people to deal with declining food yields. However, when agricultural practices are sustainably coded, it can help reduce the number of natural resources used, for example, by increasing water quality and soil fertility. Entrepreneurship is correlated with accelerating the growth and prosperity of an economy (Tiwari, 2017). Entrepreneurial activities have long been playing an important part in human development and progressiveness and the growing plethora of problems have forced farmers to learn new set of capabilities in order to sustain in the competitive market. As entrepreneurship drives innovation and employment, we see a recent rise in the growth of the sustainable movement in agriculture. Agricultural entrepreneurship is highly on the rise and it includes both farm activities and non-farm activities as well. Rural livelihoods are dependent directly or indirectly on agriculture, backyard animal husbandry, and small-scale fisheries. Entrepreneurship ventures taken on this is a means to diversify income generation activities of the rural communities (Kumari and Kumar, n.d.). Some of the entrepreneurial skills that agripreneurs have showcased include executing product specialisation, market orientation, enterprise diversity, up scaling the product, process innovation, product development and also one important feature of agripreneurship is the existence of several homogenous groups. Consequently, given the constant uncertainty we face, it is imperative that we accelerate the adoption of sustainable practices as agriculture is intertwined with employment, food system, and the overall functioning of the human population as a whole. With this background, agripreneurship is viewed as a solution to address these difficulties. It is clear from the amount of literature already in existence that engaging in agribusiness activities has improved the lives of the individuals by lowering their vulnerability and boosting their available resource, which is thus, the foundation of sustainability.

3.3 Revolution Brought by Sustainable Agricultural Entrepreneurship

Entrepreneurship drives innovation, job creation, and economic growth and empowers individuals to take charge of their situation to build better future for themselves while sustainability denotes a person's capacity to provide for themselves in the long run. Entrepreneurship and sustainability are two powerful forces that can create a positive impact. This intersection between entrepreneurship and sustainability is becoming increasingly important as we face uncertainties, risks, and challenges. But farmers who have been practicing traditional methods can be quite heterogenous in their approach towards agriculture leading to potentially environmentally harmful practices. Navigating these thin lines and trying to make them adopt newer, sustainable techniques is a daunting task. But modern, newer problems demand creative solutions which we find in the existing literature on the changes brought by the promotion or practices of sustainable agriculture. One of the most common themes found was the agricultural practices of the entrepreneurs aligning with sustainable theories; particularly that of the Sustainable Livelihood theory by the Department for International Development or the DFID under the United Kingdom's government and the recent Sustainable Development Goals or the SDG-17 by the United Nations. We also found that agricultural entrepreneurship studies were mostly in the rural context, with destinataires of the study being women farmers, organic farmers, farmers with food security issues, youth farmers or students belonging to agricultural universities (Dias et al., 2018). Studies showed farmers adopting modern technology (Dahiya et al., 2023; Meludu, 2020), mitigating climate change through the adoption of organic farming which effectively led to have enhanced the soil quality of the farmers (Abdalla & Chen, 2021; Nzama, 2021; Opiyo et al., 2023; Verner et al., 2022), promoting innovative agricultural practices at grassroots level (Pandey & Kamle, 2023; Sadasivan et al., 2023; Tiwari & Nakra, 2022), perceived increase in employment generation (Dixit & Sakunia, 2022; Guha et al., 2023), transformation of subsistence agriculture (Chakravarty et al., 2021), effective utilisation of resources (Wu et al., 2022), intervention through the use of modern technology (Dahiya et al., 2023; Meludu, 2020), development of the rural communities (Anil & Misra, 2023), plurification or diversification of farm activities for increase in income of the farmers and effective utilisation of the farm resources leading to a more robust and resilient livelihood for the participants (Kolawole et al., 2023; Little et al., 2023) and addressing issues of food security (Brown et al., 2022). The agricultural sector is facing several crises such as changing consumer preferences, population boom, environmental degradation, and increase in demand for food supply. Therefore, we find that intervention and proper training can enable entrepreneurs in agriculture to be proactive leading improvement of their own well-being but also through contribution to food security, rural employment, and economic progress.

3.4 Innovation in Sustainable Agricultural Entrepreneurship

The agricultural sector is a major source of economic activities around the world which is also laden with clumsiness and risks. In order to mitigate the risks and to remain viable and sustainable in the long run to meet the demands of the global population, several innovative practices have come up which are paving a new direction towards how agriculture is usually practised in a traditional manner. *Innovative practices* that are climate-friendly can be instilled and demonstrated among farmers and students who are studying agriculture or vocational programmes and looking for taking up agriculture in the long run (Musa et al., 2023). Practices such as imparting knowledge on food nutrients, seasonal calendar, management of soil and water health, diseases in plants, use of certain organic fertilisers, conservation agriculture-based sustainable intensification or CASI which is a host of sustainability practices that can prove to be profitable and productive for farmers, inclusive innovation and precision farming with the help of using drones, satellite imagery, and automation tools can be promoted as a crucial part of sustainability practices at the production level (Brown et al., 2022). *Social groups* which include extending support for formation of farmer's associations, producer's organisation and cooperatives, etc., besides helping the farmers in production enhancement and marketing to fetch better returns have grown among farming communities at the grassroots level. Social organisations support the creation of novel ideas, possibilities for construction, opportunities for collaboration, and access to financial resources, all of which contribute to a better standard of living (Muralikrishnan & Singh, 2018). *Diversification of activities* among farmers has become popular among farmers where it seeks to enhance the source of income using the available resource base. Usually, farm diversification activities are classified into off-farm, non-farm, and off-farm activities. Agro-climatic conditions, a lack of farmland, declining livestock productivity, poor infrastructure, a lack of financial services, a dearth of capital, inadequate access to credit services, little to no access to markets, lack of financial services, lack of skill and experience, inadequate skill training, lack of technical backing and awareness, etc., are some of the problems that prevent livelihood diversification activities (Wondim, 2019). Agricultural products suffer huge losses when post-yield supply chain lacks and fails deliverance to markets on time (Priyadarshi, 2024). Therefore, *adding value* to agricultural yields is fast becoming the desired way to minimise losses faced by farmers; even linked with increasing the net income of farmers to those of farms without value processing units (Gowda et al., 2023; Ngozi & Feleke, 2023). There have been efforts in recent times by industry experts and researchers alike to bring focus on generating revenue through adding value to agri-products and minimising post-harvest losses and sustain production of food through several participatory efforts. Although, there should be a heightened push to bridge agricultural industries, markets, and farms to promote value addition, especially among farmers to a broader facet. Agriculture has progressed from being a mere provider of raw materials to becoming a full-blown industry where entrepreneurship development agencies are taking part in promoting market awareness for farm entrepreneurs. Through the usage and blend of traditional knowledge and digital tools, farmers are now gaining access to information, optimising resources and gaining market awareness (Bei et al., 2024). *Marketing in agriculture* is considered a key driver for growth in agricultural entrepreneurship where innovation, branding, and product development lead to business growth. Bridging the gap between academia and field can enable farmers to bring sustainable innovation from the available resources (Chakraborty et al., 2023). *Organic farming* promotes farm practices without the use of synthetic fertilisers or limited usage of synthetic inputs thereof, which can have a potentially harming effect on the environment. It is one of the oldest forms of farming and has gain traction led by an appeal towards preserving soil health and subsequent effects brought on by modern farm practices. The US,

European Union are big markets for organic products if the quality of the produce adheres to their requirements. Not only that but newer generations are becoming more aware which is leading for the demand of organic products. However, it should also be noted that some studies have pointed out that the practice of organic farming requires prolonged commitment as retailers tend to influence the type of inputs utilised by the farmers for farming which can only be tackled by intensifying advanced knowledge on use of certain inputs in farms (Fan et al., 2024). *Precision or vertical farming* is an emerging technique that has gained practice for its efficient utilisation of space in urban areas. This method requires little support of climatic conditions as seen with rain-fed agricultural systems and immunity from ecological crisis and is promoted as a solution for food security and improving household health (Biancone et al., 2022; Gugganig, 2024). *Hydroponics* is another inventive method where plants are grown in water which consists of nutrients required for the growth of the plants rather than the usage of soil. Several innovations in this context such as the use of smart technologies has enabled people to start growing food at home (Angeloni, 2020). Agricultural by-products and wastes are often not utilised properly. This has presented an opportunity for experts to transform it into something that is economically viable for financial gains. Studies have found *bioenergy* production has led to improved environmental condition and leading to economic gains (Shamsuddoha et al., 2024). *Greenhouse farming* is a newer and cleaner method of agriculture as opposed to the subsistent way of farming in an environmentally sustainable way. This method can ensure sustained supply of agri-products throughout (Forkuor et al., 2022).

4 Competitive Advantage

Subsistent practices in agriculture can be short-term and not economically achievable for longer duration as this can have a negative impact on the environment likely by degrading the soil quality, reduction in the water retaining capacity of the soil, and decreasing the production level of the crops. Therefore, sustainable agripreneurship is seen as a superior alternative to achieve sustainability in various sectors of agriculture. This sector employs nearly 1 billion people and accounts for 1.3 trillion dollars' worth in food production annually (World Wildlife Fund n.d.), but there is a crippling need to increase food production by 70% if the food demands need to satisfied by 2050 (Aznar-Sánchez et al., 2020). It was found that nearly one-third of the world's farms are practising or incorporating some forms of sustainable agricultural practices such as organic farming and other forms of sustainable intensification activities that were found to reduce negative impacts of inputs like pesticides, degradation in soil quality, and carbon emissions (Washington State University, 2018). For example, *greenhouse farming* is seen as an alternative to ensure food supply where it was found that commercial greenhouse farming was worth at 34.8 billion dollars in 2021 and is estimated to grow to 78.9 billion dollars by 2030 because of its effective utilisation of space and impartiality towards weather conditions

(Poulsen, 2023). Hydroponics farming is also growing as one of the climate-conscious ways of farming and is seen as a strong potential the threats faced by the current practices of agriculture. This method uses 90% less water and allows the growing capacity of up to 3–10 times, can efficiently isolate diseased or dying crops and can lead up to 7–14 times more crop growth cycles than the usual method (Boylan, 2020). Organic farming is also emerging as the demand for organic products are increasing. It is currently being practised in 191 countries and cultivated by at least 4.5 million farmers (FiBL, 2023). It is evident that in order to meet the increasing food demands, relying solely on traditional practices will not be enough as it also impacts the environment negatively. Therefore, incorporating sustainable practices, particularly in the field of agriculture is found to be feasible in the long run. For one, it increases production especially for small-scale farmers who can expect increase in crop production cycle, efficiently uses environmental resources, reduces the harmful effects of pests and crop diseases, increases yield in production, and improves the soil fertility and increases nitrogen content in the soil (Mousavi & Eskandari, 2011). Not only that, sustainable agricultural entrepreneurship is seen as an elevation from traditional practices providing employment, enhancing the traditional knowledge and skills, and acting as the link among human capital, environment, and the economy. This will therefore, not only boost food production in the coming decades but will make farmers aware of the benefits derived from sustainable practices in agriculture entrepreneurship in the long run.

5 Conclusion

Sustainability in agricultural entrepreneurship are practices that are viable in the long term and which are not detrimental to the natural resource base. As studied by many scholars, food security is becoming an issue for many countries and this is putting more pressure on the production capacity of agriculture. The world's food demand will need to be increased to 70% by 2050 to keep up with the demands of the increasing population. Subsistence agricultural practices have had many negative impacts on humans and the environment and adversely affecting the soil, air, and water quality. Therefore, for sustenance of livelihood and biodiversity both now and in the future, sustainable agricultural entrepreneurship is a solution for bringing this balance and harmony with the ecosystem. We find some innovative practices that are being utilised and incorporated by farmers through the aid of training and development such as vertical farming, hydroponics, greenhouse farming, and organic farming; all developed around one central theme: ensuring food security and food production without undermining the natural resource base. There are to some extent, some challenges to implementation of innovative sustainable farm practices such as expenses pertaining to the technology required to adopt certain farm practices, technical know-how, farm mechanisation, indifference towards adoption of sustainable practices, lack of support or any policies from the responsible authorities, lack of monitoring and evaluation, gap between academia and practice and in general, the lack of awareness among the farm entrepreneurs. However, since the overall positives outline the negative aspects, one should look at the prospects arising from best sustainable practices in the long run. Agricultural entrepreneurship therefore, will bridge the gap between the economy and biodiversity.

References

Abdalla, T. A. O., & Chen, Y. (2021). A case study on the opportunity and challenges faced by organic farming in Jiangsu Province. *Agroecology and Sustainable Food Systems, 45*, 1327–1374. https://doi.org/10.1080/21683565.2021.1932684

Afroz, S., Singh, R., Nain, M. S., & Mishra, J. R. (2022). Determinants for Agripreneurship development under Agriclinics and agribusiness centers (ACABC) scheme. *Indian Journal of Agricultural Sciences, 92*, 258–262. https://doi.org/10.56093/ijas.v92i2.122248

Agu, P. (2020). Determinants of household income and employment choices in the rural Agripreneurship economy. *Studies in Agricultural Economics, 122*, 96–103. https://doi.org/10.7896/j.2044

Angeloni, S. (2020). Domo Farm 4.0. *International Journal of Grid and Utility Computing, 11*, 135–142. https://doi.org/10.1504/IJGUC.2020.105546

Anil, K., & Misra, A. (2023). Bakrichhap agro tourism and natural products private limited (BATNPPL): A case for community based entrepreneurship promoting integrated rural development. *Emerald Emerging Markets Case Studies, 13*, 1–28. https://doi.org/10.1108/EEMCS-12-2022-0512

Aznar-Sánchez, J. A., Velasco-Muñoz, J. F., López-Felices, B., & Román-Sánchez, I. M. (2020). An analysis of global research trends on greenhouse technology: Towards a sustainable Agriculture. *International Journal of Environmental Research and Public Health, 17*, 1–22. https://doi.org/10.3390/ijerph17020664

Bei, N., Li, W., & Cao, L. (2024). Digital integration and entrepreneurial success: Examining causation and effectuation in rural South China. *Journal of the Knowledge Economy*. https://doi.org/10.1007/s13132-024-01870-8

Biancone, P. P., Brescia, V., Lanzalonga, F., & Alam, G. M. (2022). Using Bibliometric analysis to map innovative business models for vertical farm entrepreneurs. *British Food Journal, 124*, 2239–2261. https://doi.org/10.1108/BFJ-08-2021-0904

Boylan, C. (2020). The future of farming: Hydroponics. Princeton Student Climate Initiative. Accessed June 5, 2024, from https://psci.princeton.edu/tips/2020/11/9/the-future-of-farming-hydroponics

Brahma, M., & Tripathi, S. S. (2020). Indian society of agribusiness professionals: Helping Agripreneurs innovate. *Emerging Economies Cases Journal, 2*, 15–23. https://doi.org/10.1177/2516604220920752

Brown, P. R., Anwar, M., Hossain, M. S., Islam, R., Siddquie, M. N. E. A., Rashid, M. M., Datt, R., Kumar, R., Kumar, S., Pradhan, K., Das, K. K., Dhar, T., Bhattacharya, P. M., Sapkota, B., Thapa Magar, D. B., Adhikari, S. P., Rola-Rubzen, M. F., Murray-Prior, R., Cummins, J., Maharjan, S., Gathala, M. K., Brown, B., & Tiwari, T. P. (2022). Application of innovation platforms to catalyse adoption of conservation Agriculture practices in South Asia. *International Journal of Agricultural Sustainability, 40*, 497–520. https://doi.org/10.1080/14735903.2021.1945853

Caradonna, J. L. (2014). *Sustainability: A history*. Oxford University Press.

Carr, S., & Roulin, A. (2016). An exploration of agripreneurship scope, actors and prospects. *Agenda, 2063*, 16.

Chakraborty, M., Parida, P., Mandal, N. K., Patnaik, S., & Chakraborty, S. (2023). Marketing growth dimensions of Agri-entrepreneurship: An empirical study in Ganjam District of Odisha,

India. *Universal Journal of Agricultural Research, 11*, 1013–1021. https://doi.org/10.13189/ujar.2023.110608
Chakravarty, A., Whitbread, A., Kaur, P., Selvaraj, A., Mazumdar, S. D., Philroy, J., Durgalla, P., Mane, H., & Sharma, K. K. (2021). Benefitting smallholder farmers in Africa: Role of ICRISAT. In *International political economy series* (pp. 73–93). https://doi.org/10.1007/978-3-030-54112-5_4
Chand, K. K. (2019). Agripreneurship: A tool for economic development of India in the new millennium. *International Journal on Recent Trends in Business and Tourism, 3*, 19–25.
Dabson, K., & Markley, D. (2010). Agripreneurs: A free range of opportunity.
Dahiya, K., Taneja, S., & Özen, E. (2023). To analyse the impact of multi-media technology on the rural entrepreneurship development. In *Contemporary studies of risks in emerging technology, Part A* (pp. 221–240). https://doi.org/10.1108/978-1-80455-562-020231015
De Haan, L. J. (2000). Globalization, localization and sustainable livelihood. *Sociologia Ruralis, 40*, 339–365. https://doi.org/10.1111/1467-9523.00152
Dhingra, G., Arora, N., & Panday, P. (2018). Role of NGOs in the promotion of Agriculture. *Journal of Global Economy, 14*, 91–99. https://doi.org/10.1956/jge.v14i4.500
Dias, C. S. L., Rodrigues, R. G., & Ferreira, J. J. (2018). What's new in the research on agricultural entrepreneurship? *Journal of Rural Studies, 65*, 1–17. https://doi.org/10.1016/j.jrurstud.2018.11.003
Dias, C. S. L., Rodrigues, R. G., & Ferreira, J. J. (2019). Agricultural entrepreneurship: Going back to the basics. *Journal of Rural Studies, 70*, 125–138. https://doi.org/10.1016/j.jrurstud.2019.06.001
Dixit, K., & Sakunia, D. (2022). Start-up village entrepreneurship Programme: 'From local to vocal. *Contemporary Voice of Dalit, 15*(2), 216–229. https://doi.org/10.1177/2455328X211042723
Dollinger, M. J. (2003). *Entrepreneurship: Strategies and resources*. Pearson International Edition.
Washington State University. (2018). Environmentally friendly farming practices used by nearly 1/3 of world's farms. *WSU Insider*. Accessed June 5, 2024, from https://news.wsu.edu/press-release/2018/08/27/environmentally-friendly-farming-practices-now-nearly-33-worlds-farms/
Falola, A., Mukaila, R., Akanbi, S.-u. O., Ahmad, S., & Christianah, O. (2022). Agripreneurial drive among women Shea butter processors in Kwara state, Nigeria: Motivating factors and efficiency. *Journal of Agriculture and Rural Development in the Tropics and Subtropics, 123*, 175–187. https://doi.org/10.17170/kobra-202210116962
Fan, X., Wang, Z., & Wang, Y. (2024). Rural business environments, information channels, and farmers' pesticide utilization behavior: A grounded theory analysis in Hainan Province, China. *Agriculture (Switzerland), 14*, 196. https://doi.org/10.3390/agriculture14020196
FiBL. (2023). The world of organic Agriculture 2023: Statistics and emerging trends. FiBL. Accessed June 5, 2024, from https://www.fibl.org/en/shop-en/1254-organic-world-2023
Forkuor, G., Amponsah, W., Oteng-Darko, P., & Osei, G. (2022). Safeguarding food security through large-scale adoption of agricultural production technologies: The case of greenhouse farming in Ghana. *Cleaner Engineering and Technology, 6*, 100384. https://doi.org/10.1016/j.clet.2021.100384
Gowda, M. J. C., Rana, R. K., Pal, P. P., Dubey, S. K., Kumar, A., Meena, M. S., Singh, R., Bordoloi, R., Bhaskaran, A., Raut, A. A., Rajesh, T., Kumar, B., & Thimmapa, K. (2023). Economic performance of enterprises promoted under ARYA and relationship with entrepreneurial competencies. *Indian Journal of Extension Education, 59*, 10–15. https://doi.org/10.48165/IJEE.2023.59203
Gugganig, M. (2024). Vanguard visions of vertical farming: Envisaging and contesting an emerging food production system. *Science Technology and Human Values*. https://doi.org/10.1177/01622439241240796
Guha, S., Mandal, A., Kujur, F., & Poddar, S. (2023). Creating employment and sustainable livelihood through entrepreneurship: A study of artisans. In *Multidisciplinary approaches to sustainable human development* (pp. 26–50). IGI Global. https://doi.org/10.4018/978-1-6684-8223-0.ch002

Kabir, M. S., Hou, X., Akhter, R., Wang, J., & Wang, L. (2012). Impact of small entrepreneurship on sustainable livelihood assets of rural impact of small entrepreneurship on sustainable livelihood assets of rural poor women in Bangladesh. *International Journal of Economics and Finance, 4*, 265–280. https://doi.org/10.5539/ijef.v4n3p265

Kadirvel, G., Banerjee, B. B., & Assumi, S. R. (2020). Agri-business in North East India: Current status, potential ventures and strategies. *Current Journal of Applied Science and Technology, 39*, 74–85. https://doi.org/10.9734/cjast/2020/v39i3331021

Kolawole, O. D., Hambira, W. L., & Gondo, R. (2023). Agrotourism as peripheral and Ultraperipheral community livelihoods diversification strategy: Insights from the Okavango Delta, Botswana. *Journal of Arid Environments, 212*, 104960. https://doi.org/10.1016/j.jaridenv.2023.104960

Kumari, P., & Kumar, T. (n.d.). Opportunities, roles and prospect challenges of rural entrepreneurs in economic development in India. *Dimensions of Agricultural Science*, 354–364.

Lalhmingthanga, R., Singh, R. K., & Singh, E. N. (2020). Agripreneurship in organic crops: Challenges and opportunities in Mizoram. *Mizo Studies, 9*, 543.

Little, V. J., Ho, H. H. P., & Eti-Tofinga, B. (2023). Not WEIRD at all! Towards more pluralistic economies and sustainable livelihoods. *Journal of Macromarketing, 43*, 190–214. https://doi.org/10.1177/02761467231157429

Magagula, B., Tsvakirai, C. Z., & Francis Group. (2020). Development in practice youth perceptions of Agriculture: Influence of cognitive processes on participation in Agripreneurship. *Development in Practice, 30*, 234–243. https://doi.org/10.1080/09614524.2019.1670138

Meludu, N. T. (2020). Showcasing creativity in home economics through utilization of social Media for Entrepreneurship Development. *Journal of Home Economics Research, 27*, 31–41.

Mousavi, S. R., & Eskandari, H. (2011). A general overview on intercropping and its advantages in sustainable Agriculture. *Journal of Applied Environmental and Biological Sciences, 1*, 482–486.

Muralikrishnan, L., & Singh, R. (2018). Social capital based agripreneurship models for inclusive and sustainable development in India social value system development about judicious use of pesticides and other agro-chemicals and entrepreneurship among youth view project development of innovative. *Trends in Biosciences, 11*, 1–7.

Musa, P. D., Fatimahwati, S., Idris, P. H., & Rozaidah, P. S. (2023). Exploring the concept of entrepreneurial identity in youth Agripreneur program. *International Journal of Training Research, 21*, 211–225. https://doi.org/10.1080/14480220.2023.2194668

Ngozi, S., & Feleke, S. (2023). Enhancing youth's retention in potato farming enterprise through consortium agribusiness models in Tanzania. *Crop Science, 64*(3), 1421–1431. https://doi.org/10.1002/csc2.21055

Nzama, A. T. (2021). Tackling climate change through craft development: The case of rural women in UPhongolo local municipality. *Jamba: Journal of Disaster Risk Studies, 13*, 1–10. https://doi.org/10.4102/JAMBA.V13I1.1140

Opiyo, S. B., Opinde, G., & Letema, S. (2023). A perspective of sustainable livelihood framework in analysis of sustainability of rural community livelihoods: Evidence from Migori River watershed in Kenya. *International Journal of River Basin Management, 22*(4), 627–643. https://doi.org/10.1080/15715124.2023.2216019

Pandey, N., & Kamle, D. (2023). Building start-up ecosystem through social Enterprise business model: A study reference to women empowerment by social enterprises. In *Springer proceedings in business and economics* (pp. 579–591). https://doi.org/10.1007/978-981-99-0197-5_36

Parambil, G. D. (2016). Agriprenuership: A conceptual framework of agricultural entrepreneurship in India. *Bharat Mata Journal of Multidisciplinary Studies, III*, 64.

Poulsen, T. (2023). Driving the greenhouse industry to sustainable growth. Agritech Tomorrow. Accessed June 5, 2024, from https://www.agritechtomorrow.com/story/2023/02/driving-the-greenhouse-industry-to-sustainable-growth-/14375/

Priyadarshi, R. (2024). Observation of post-yield supply chain impediments for spoilage mitigation and revenue generation opportunities at countryside. *Journal of Global Operations and Strategic Sourcing, 17*, 127–145. https://doi.org/10.1108/JGOSS-06-2023-0052

Sadasivan, E., Manoharan, K., Das, M., & Bhattacharya, S. (2023). Rural employment through product development: Entrepreneurial framework for Grassroot level innovations. *Journal of the Knowledge Economy, 14*, 4807–4827. https://doi.org/10.1007/s13132-022-01071-1

Shamsuddoha, M., Koul, S., & Taylor, I. W. (2024). Eco-efficient dairy waste treatment: Validating a sustainable system dynamics framework. In *Operations research forum* (Vol. 5). Springer International Publishing. https://doi.org/10.1007/s43069-023-00290-9

Tiwari, N. (2017). Women entrepreneurship in India: A literature review. *Amity Journal of Entrepreneurship, 2*, 47–60.

Tiwari, M., & Nakra, A. (2022). Jute for life: Reconciling the social and business goals through sustainable entrepreneurship. *Emerald Emerging Markets Case Studies, 12*, 1–34. https://doi.org/10.1108/EEMCS-03-2021-0083

Uplaonkar, S. S., & Biradar, S. S. (2015). Development of Agriculture in India through Agripreneurs. *International Journal of Applied Research, 1*, 1063–1066.

Verner, D., Vellani, S., Goodman, E., & Love, D. C. (2022). Frontier agriculture: Climate-smart and water-saving agriculture technologies for livelihoods and food security. In *New forms of urban agriculture: An urban ecology perspective* (pp. 159–186). Springer Nature Singapore. https://doi.org/10.1007/978-981-16-3738-4_9

Vidani, J., Jagruti, L., Institutes, K., & Plaha, N. (2017). Agripreneurship: A reincarnation of Indian agricultural sector. In *International Conference on enhancing Economic Productivity and Competitiveness through Financial and Monetary Reforms* (pp. 154–159).

Wondim, A. K. (2019). Determinants and challenges of rural livelihood diversification in Ethiopia: Qualitative review. *Journal of Agricultural Extension and Rural Development, 11*, 17–24. https://doi.org/10.5897/jaerd2018.0979

World Bank. Agriculture and food. (2024). Accessed May 27, 2024, from https://www.worldbank.org/en/topic/agriculture/overview

World Wildlife Fund. Sustainable agriculture. (n.d.). Accessed May 26, 2024, from https://www.worldwildlife.org/industries/sustainable-agriculture

Wu, B., Geng, B., Wang, Y., Scott McCabe, L., Liao, L. Z., & Deng, B. (2022). Reverse entrepreneurship and integration in poor areas of China: Case studies of tourism entrepreneurship in Ganzi Tibetan region of Sichuan. *Journal of Rural Studies, 96*, 358–368. https://doi.org/10.1016/j.jrurstud.2022.11.012

Yoganandan, G., Aziz, A., Abdul Rahman, M., & Vasan, and Abdelrhman Meero. (2022). Evaluating Agripreneurs' satisfaction: Exploring the effect of demographics and Emporographics. *Journal of Innovation and Entrepreneurship, 11*, 2–22. https://doi.org/10.1186/s13731-022-00193-9

Zhang, C., Fang, Y., Chen, X., & Congshan, T. (2019). Bibliometric analysis of trends in global sustainable livelihood research. *Sustainability, MDPI, 11*, 1–28. https://doi.org/10.3390/su11041150

The Impact of Environmental, Social, and Governance (ESG) Objectives' Effect on Investment Decisions: Retail Investors' Knowledge and Views on Sector-Specific ESG Factors

Deep Bhowmick, Harshal Anil Salunkhe, and Mohit Boralkar

Abstract Following the global financial crisis, Environmental, Social, and Governance discretion in stock market investment decisions has become necessary for sustainable development. Environmental, Social, and Governance consider investing qualities such as accountability, risk minimization, and long-term return. This study investigates how individual stock market investors perceive and are aware of Environmental, Social, and Governance concerns, as well as how the purpose of an investment influences how judgments are made regarding it. The study's sequential mixed research methodology begins with an interview and progresses to a cluster and snowball sampling procedure for a survey of stock market investors. The study's findings show how the Environmental, Social, and Governance factor moderates investment decisions, with implications for managers, investors, and regulators. Investors will be assured a long-term return on investment as well as accountability to the environment, society, and economic equilibrium, which could lead to sustainable development. It is also recommended that future research be performed in various states and countries with diverse causes and conceptions.

1 Introduction

ESG investing has evolved to address the needs of institutional, retail, and some government investors who want to better integrate long-term financial risk and opportunity into their investment decisions to create long-term value. Long-term

D. Bhowmick · M. Boralkar (✉)
Christ (Deemed to be) University, Bangalore, India
e-mail: mohit.boralkar@christuniversity.in

H. A. Salunkhe
School of Business, MIT World Peace University, Pune, India

A. Sharma et al. (eds.), *The Synergistic Effect of Sustainable Business Practices on Corporate Performance*, https://doi.org/10.1007/978-981-96-3375-3_4

challenges in ESG categories include controversies and adverse risks with the potential to reduce equity value over time and increase credit risk. As a result, it aims to combine improved risk management with greater portfolio returns and include investor and benefit values into investment techniques. Investing early in sustainable economic growth can help vulnerable communities mitigate the effects of climate change. Private sector investments that assist vulnerable people and communities are adapting to the repercussions of climate change, particularly those that facilitate better the use of increasingly scarce resources or aid in their renewal and restoration, are an essential component of the broader sustainable economic paradigm. The inextricable links between human and environmental well-being, social and financial stability, and long-term profitability of the private sector lay the groundwork for green economic growth.

The start of the COVID-19 pandemic in the first quarter of 2020 caused a sharp and rapid drop in global markets (KHIN et al., 2018). After the devastating pandemic, ESG was touted as a "resilience factor" by "responsible investing" investment firms and ESG data providers. (Morningstar, for example, referred to ESG as an "equity vaccination" against the "pandemic-induced pullback" in the stock market, which dropped 34% from its January 19, 2020 peak to March 23 before recovering slightly toward the end of Q1 2020.) Similar arguments can be made for the whole COVID year of 2020; after the initial hype around ESG as a hedge against "opportunity risk," it was no surprise to see record inflows into sustainable funds into Q1 2020. According to CNBC's report, demand grew yearly. Even while investors were very enthusiastic about ESG-tilted investments, by mid-2020, there was doubt about whether ESG was a reliable way to protect profits during a crisis. Opponents said that the concept of ESG scores was outdated and that the period for "blank aggregation" had expired at the beginning of 2021.

Choosing a stock to invest in is a complex process that entails looking through all the options available on major stock exchanges. Standard economic theory holds that people are rational actors who may seize opportunities by forming unbiased opinions based on their expectations, experiences, and knowledge. However, the behavioral paradigm of financial decision-making—that is, how investors perceive the world and make investment decisions—describes how investors perceive the world as well as make investment decisions (Cohen & Kudryavtsev, 2012; Bikas et al., 2013). Previously, investment decisions were followed by an ordinary triangle that covered risk, liquidity, and return; however, an increasing number of investors now use the phenomenal square, which covers liquidity, risk, return, and sustainability (von Wallis & Klein, 2014). Investors' investment decision-making processes are no longer uniform across the board but vary by investor. Diverse investors utilize distinct approaches when selecting equities or bonds. One set of investors could base their investment choice on the financial results of the venture.

In contrast, other investors may view economic consequences and ESG issues in their investment decisions. The current study investigates stock market investors' ESG investment decisions by measuring their readiness to pay a premium price at the expense of return and the percentage of asset allocation investments in companies to rigorous ESG practices (Sultana et al., 2018). Based on the preceding study, the following hypothesis has been formulated:

H_1: The investor perception and various industries' ESG practices are significantly correlated.

There are numerous determinants and intended results for corporate ESG adoption. Initially, management can view ESG as a strategic choice. The efforts of a firm in sustainability can enhance firm value and make sustainability, bring cost leadership and differentiation savings, have a positive effect on employees' work engagement (Agarwal et al., 2012), and improve employee productive behavior (Park, 2020) as well as customer loyalty (Kim & Park, 2017).

Second, ESG may support opportunity and risk management to benefit management, employees, critical suppliers, and clients. Corruption threatens the survival of new businesses (Nam et al., 2020) and it may significantly affect profitability and stock prices (Thakur et al., 2019). Trying to deal with threats could be an assertive way of ensuring a company's long-term viability (Jo & Na, 2012). Risk management should also include the domain of the reputation of the business. More than actual offenses, news disclosures about illegal activity have hurt the company's income (Song & Han, 2015). De Franco (2019) studied the impact of ESG-related issues on stock and portfolio performance and discovered that low-ESG or large ESG-related portfolios underperformed. According to recent research, highly sustainable businesses face fewer adverse risks and are more resilient to volatility. Participating in ESG reduces a company's downside risk, according to Hoepner et al. (2016). Ilhan et al. (2018) state that companies with poor ESG profiles caused by excessive carbon emissions face a higher tail risk. According to studies, financial and non-U.S. corporations with high ESG ratings outperformed others during the global financial crisis (Cornett et al., 2016).

Third, a business must use ESG to comply with legal or regulatory standards because it owes it to society. While ESG disclosures alone will not make a business entity ethical, they will significantly increase its ethical selection when making pertinent judgments. By reviewing company practices, ESG financial companies and other financial lending institutions may be regarded as change agents (Escrig-Olmedo et al., 2019). Finally, corporate ESG implementation may benefit the firm's stakeholders. Institutional investors have become much more engaged in ESG investing for two reasons. To begin with, ESG investing promotes ethical investment practices. Second, ESG investments are seen as a way to improve managed portfolio performance, increase returns, and reduce portfolio risk (Broadstock et al., 2020). Shareholders anticipate the company's economic performance and long-term viability in the future. Verga Matos et al., (2020) found that sustainable businesses were more likely to pay consistent dividends. These stakeholders, including shareholders, benefit from stronger long-term relationships with sustainable businesses. Several pieces of literature suggest that investing in companies implementing strong ESG practices yields a higher return on investment and that outstanding ESG management throughout corporations can proactively avoid unexpected cash flow shocks (Lee et al., 2013). Friede et al. (2015) reviewed ESG/ Socially responsible investing (SRI) research and discovered a strong positive correlation between ESG and financial success.

1.1 Environmental, Social, and Governance (ESG) Issues

Investing in ESG issues requires an examination of a company's effect on the environment (e.g., carbon emission levels, water and energy use), community (e.g., product safety, safety and health, philanthropy, and reasonable principles), and governance practices reliability (e.g., bribery and corruption, board independence, and stakeholders' activity and security (Sultana et al., 2018). The ESG portfolio manager examines all these characteristics to establish a diversified portfolio, which has become a key indicator of managerial skills, risk assessment, and non-financial performance. The Association for Sustainable and Responsible Investment in Asia reports that since 2011, the value of sustainable investments has increased by 22% annually. The ESG problems is Asia most widely used sustainable investing technique. As a result, ESG problems in stock market investment have gained prominence worldwide (Sultana et al., 2018).

H_2: There is a considerable association between awareness of ESG and sectors.

The functioning and efficiency of biological systems and the environment are the subjects of environmental concerns. A few examples are the loss of biodiversity, changes to the nitrogen and phosphorus cycles, waste management, climate change, ozone layer depletion, energy efficiency, renewable energy sources, and air, water, or resource contamination. Environmental concerns are one of the most important considerations when choosing an investment. For instance, previous research conducted in Japan, India, France, and Australia has corroborated this (Sreekumar Nair & Ladha, 2014). Environmental concerns were the most important considerations for Indian investors when making non-economic investment decisions.

H3: The industry sector and ESG issues are significantly correlated.

Human rights, dangerous weapons, consumer protection, enslavement, child labor, and bonded labor are examples of social concerns. Other topics include supply chain labor standards, diversity, freedom of expression and association, health and access to medicine, employee engagement and people management, community relationships, and conflict resolution. Shareholders in Australian investment companies consider social issues, community–employee engagement, and human rights when making an investment decision (De Zwaan et al., 2013); furthermore, Australian investors prioritize social concerns over other problems like the environment and government. Moreover, Australian investors value social concerns more than other problems, such as the climate and government (Pérez-Gladish et al., 2012). However, stock market investors in Brazil do not consider social issues (Miralles-Quirós et al., 2018). An increasing number of investors worldwide are evaluating the social concerns related to the companies they invest in.

Investor economic rationality is strongly linked to corporate governance procedures. Profitability is essential for resource allocation, long-term company viability, and the overall economy (Man-Kit, 2015). Businesses with effective governance practices are more likely to participate in ethical, social, and environmental projects. The size of the board, structure, diversity, skill sets, independence, internal control systems and risk management, executive compensation, information disclosure, business practices, shareholder rights, stakeholder interaction, the relationship between a company's management staff and stakeholders, bribery, and corruption are among the governance issues that worry businesses and other investment entities. Investors prefer social problems as long as they are accompanied by high financial returns, suggesting a preference for the company's governance (Rakotamavo, 2011, Pérez-Gladish et al., 2012). In Australia, 64% of respondents support corporate governance when making an investment decision (De Zwaan et al., 2013); however, stock market investors in Brazil do not consider governance issues (Miralles-Quirós et al., 2018). Due to the Great Recession, several stakeholder groups have expressed a need for more robust governance. However, in a country where investors have just witnessed a stock market crisis and issues with corporate governance, it is unclear if stock markets are taking corporate governance concerns into account.

The investment horizon, which encompasses short-, mid-, and long-term investment strategies, is essential for making ESG investments. The significance of ESG investments and financial sector participation in the investment horizon should be prioritized (Carolina Rezende de Carvalho Ferrei et al., 2016). Since long-term strategies are associated with long-term investment gains, the investment horizon may impact ESG investment choices and value creation. Because a company's primary purpose is to maximize wealth and generate long-term value (Pavláková Dočekalová & Kocmanová, 2018), this value-creation approach of sustainability is connected to risk reduction (Alshehhi et al., 2018). Long-term investors are willing to accept a short-term drop in their investment if the return is excellent; no significant differences in the investment horizons of socially accountable and non-socially responsible shareholders were discovered among Australian investors (Pérez-Gladish et al., 2012).

In contrast, a short-term investment view is a significant hurdle to Swedish institutional investors adopting socially conscious investments (Jansson et al., 2011). On the other hand, the scope and sophistication of the specific approach and notion influence how much an investment plan promotes a long-term perspective (Man-Kit, 2015). The investment horizon may affect how risky the investment is. The importance of time horizon to traders in investment portfolios is currently being debated (Veld-Merkoulova, 2011). Analyze the importance of the investment timeframe concerning ESG matters and investment choices (Fisher & Statman, 1999), which is connected to diversification, and understand the impact of the investment horizon in aspects of ESG factors and investment strategy (Fisher & Statman, 1999). To give a comprehensive picture of sustainability practices and the company's sustainable development, further modifying and intervening factors must be looked into (Busch & Hoffmann, 2011, Schreck, 2011).

ESG risks in emerging markets are higher than in developed markets; they use responsible investing techniques to a lower extent when making emerging market investments. Addressing ESG issues, particularly in developing economies, may be highly profitable for investors aiming to minimize portfolio risk or find new investment possibilities. After reviewing the literature, I noticed that people focused more on ESG factors in developed economies and less on emerging markets. Emerging markets provide a significant investment potential and a unique mix of risks and problems in terms of environmental, social, and Governance (ESG). Some papers and articles address this topic (ESG). Still, no papers explicitly discuss investors' investment decisions in emerging markets by considering the ESG factors. Looking at the investors' point of view, this research is done to understand if ESG factors play an influential role in influencing investors' investment decisions.

2 Materials and Methods

The sequential mix research method is used in this study; an interview is conducted first, followed by a questionnaire survey. This method, helpful in investigating phenomena while creating and testing a novel instrument. The researcher gathers and analyzes qualitative data and using quantitative data gathering in the analytical phase. Initially, a semi-structured interview was conducted with academics and investors to find out their inclusive perspectives on the ESG policies of businesses in operation and the present stock market investors' thoughts on ESG matters. Second, as gathering final data was the main goal of the research, a questionnaire survey was administered to individual Indian Stock Exchange stock market investors using a combination of cluster and snowball sampling. This study employed positivist accounting research as it aims to uncover regular patterns akin to laws within empirical datasets. A single survey was distributed to both academics and individual investors in the stock market for the pilot study. After gathering the 250 responses, the survey questionnaire was adjusted to improve face validity and reliability. The study's goal and questionnaire both included ideas for interviews. For a sample size of 400, a single questionnaire was issued in total. Usable replies were examined in the end. The study then tested the hypotheses using scientific techniques. The quantitative methodologies enabled us to arrive at a logical solution to the issue brought up in the survey.

To evaluate the quantitative data, we utilized the Statistical Package for Social Science (SPSS) software, which corrects for measurement errors and produces estimates of accurate composites and components. The authors conducted an analysis employing ANOVA, chi-square, Welch, and Brown–Forsythe tests to evaluate the significance of the study and deliver the necessary interpretations.

2.1 Research Instrument

This study's thorough research instrument was modified since the long-term paradigm demands that ESG assessment scales be improved. A belief component—a mental process used to understand an object—and an emotional component—liking or hating an object—is suggested for measuring attitudes. The scales were adjusted to evaluate the three homogeneous independent variables—environmental, societal, and governance challenges. A five-point Likert scale, from "strongly disagree" to "strongly agree," was utilized. Two negative questions were included for each independent variable to prevent acquiescence and random replies. A five-point Likert scale was employed to assess both the dependent variable and the investors' investment decisions (Table 1).

Table 1 ANOVA

(I) Sector	(J) Sector ratings	Significance
Real estate	Pharmaceutical	0
	Transportation	0
	Agriculture	0
	Hospitality	0.016
	Entertainment	0
	Finance/banking/insurance	0.018
Automobile	Agriculture	0.009
	Entertainment	0.008
Information technology/software	Agriculture	0.006
	Entertainment	0.005
Pharmaceutical	Real estate	0
Transportation	Real estate	0
	FMCG	0.034
Agriculture	Real estate	0
	Automobile	0.009
	Information technology/software	0.006
	FMCG	0.002
Hospitality	Real estate	0.016
Entertainment	Real estate	0
	Automobile	0.008
	Information technology/software	0.005
	FMCG	0.002
Finance/banking/insurance	Real estate	0.018
	Transportation	0.034
	Agriculture	0.002
	Entertainment	0.002
FMCG	Transportation	0.34
	Agriculture	0.002
	Entertainment	0.002

3 Results and Discussion

The Real Estate investors positively impact the Pharmaceutical, Transportation, Agriculture, and Entertainment with a p-value of 0, Finance/Banking/Insurance with a p-value of 0.018, and hospitality with a p-value of 0.016. Given that their significance level is below 0.05, denying our theory of nothing, i.e. there is no significant relationship between investors' perception and the ESG practices of different industries. The investors working in the Automobile industry have a positive impact on the Agriculture and Entertainment industries with a p-value of 0.009 and 0.008, respectively. Since their significance level is less than 0.05, we reject our null hypothesis, i.e. there is no significant relationship between investors' perceptions and the ESG practices of different industries. The investors working in the Information Technology/Software industry positively impact the Agriculture and Entertainment sector with p-values of 0.006 and 0.005, respectively. Since their significance level is less than 0.05, we reject our null hypothesis, i.e. there is no significant relationship between investors' perceptions and the ESG practices of different industries.

The Pharmaceutical investors positively impact the Real Estate industry with a p-value of 0. Since their significance level is less than 0.05, rejecting our null hypothesis, i.e. there is no significant relationship between investors' perception and the ESG practices of different industries. The investors working in the Transportation industry positively impact the Real Estate and FMCG sectors with a p-value of 0 and 0.034, respectively. Since their significance level is less than 0.05, we reject our null hypothesis, i.e. there is no significant relationship between investors' perceptions and the ESG practices of different industries. The investors working in the Agriculture industry have a positive impact on Real Estate with a p-value of 0, Automobile with a p-value of 0.009, Information Technology/Software industry with a p-value of 0.006, and FMCG with a p-value of 0.002. Since their significance level is less than 0.05, rejecting our null hypothesis, i.e. there is no significant relationship between investors' perception and the ESG practices of different industries. The investors working in the Hospitality industry positively impact the Real Estate sector with a p-value of 0.016. Since their significance level is less than 0.05, we reject our null hypothesis, i.e. there is no significant relationship between investors' perceptions and the ESG practices of different industries. The investors working in the Entertainment industry have a positive impact on Real Estate with a p-value of 0, Automobile with a p-value of 0.008, Information Technology/Software with a p-value of 0.005, and FMCG with a p-value of 0.002. Since their level of significance is less than 0.05, rejecting our null hypothesis, i.e. there is no significant relationship between investors'perception and the ESG practices of different industries. The investors working in the Finance/Banking/Insurance industry have a positive impact on Real Estate with a p-value of 0.018, Transportation with a p-value of 0.034, and Agriculture and Entertainment with a p-value of 0.002 each. Since their level of significance is less than 0.05, rejecting our null hypothesis, i.e. there is no significant relationship between investors' perception and the ESG practices of different industries. The investors working in FMCG have a positive impact on Transportation with a p-value of 0.34, Agriculture with a p-value of 0.002, and Entertainment with a p-value of 0.002. Since their level of significance is less than 0.05, rejecting our null hypothesis, i.e. there is no significant relationship between investors' perception and the ESG practices of different industries.

The investors working in Education industry hurt Real estate, Automobile, Information Technology/Software, Pharmaceutical, Transportation, Agriculture, Hospitality, Entertainment, Finance/Banking/Insurance, and FMCG sector's ESG practices since their level of significance is more than 0.05, rejecting our alternative hypothesis, i.e. there is a positive perception of investors regarding how different sectors handles ESG practices. The null hypothesis that the variance is the same for all groups is tested using Levene's test using an F-test. A p-value of less than 05 indicates an assumption breach. Doing the non-parametric equivalent of the analysis is more acceptable in case of a breach (Table 2).

Brown–Forsythe and Welch tests are used to find significant differences among the means of more than two samples, the same as ANOVA when the assumption of homogeneity of variance is void. Both tests mentioned earlier were applied when the assumption of equal variances was not met (Table 3).

After performing the test of Homogeneity of Variances, it was seen that the Levene Statistic was significant, i.e., p-value less than 0.5, which violates the assumption, so we conducted the Welch and Brown–Forsythe tests, and the results were substantial (Fig. 1).

The Mean Plot of Sector and Sector Ratings represents the Industry Sector that handles ESG practices. From the results, investors in different sectors rated the agriculture and entertainment industries most as they took the ESG practices more efficiently and ranked real estate and FMCG the lowest. Descriptive analysis is

Table 2 Test of homogeneity of variances

		Levene Statistic	df1	df2	Sig.
Sector ratings	Based on mean	11.760	10	2718	0.000
	Based on median	8.265	10	2718	0.000
	Based on the median and with adjusted df	8.265	10	2536.955	0.000
	Based on trimmed mean	12.488	10	2718	0.000

Table 3 Test of equality of means

Sector ratings	Statistic[a]	df1	df2	Sig.
Welch	7.089	10	1086.142	0.000
Brown–Forsythe	7.123	10	2595.089	0.000

[a] Asymptotically F distributed

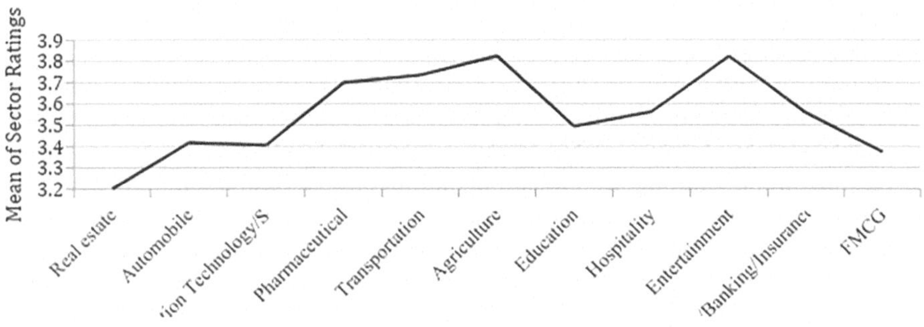

Fig. 1 Mean plot of sector ratings

Table 4 Awareness of investors of different sectors regarding ESG

Sectors	Yes	No
Real estate	55.6	44.4
Automobile	84.2	15.8
Information technology/software	100	–
Pharmaceutical	77.2	22.8
Transportation	100	–
Agriculture	100	58.8
Hospitality	41.2	38.1
Finance/banking/insurance	61.2	–
FMCG	100	–
Entertainment	100	100
Education	0	
Total	77.10%	22.90%

performed to test the awareness level of investors in different sectors, and crosstabs and chi-square tests are used to get the results (Table 4).

The awareness level of investors in the Real Estate industry is 56%, Automobile is 84.2%, Information Technology/Software is 100%, Pharmaceutical is 77.2%, Hospitality is 41.2, Finance/Banking/Insurance is 61.9%, Transportation, Agriculture, Entertainment, and FMCG industries have 100% awareness of the concept of ESG, and the investors working in Education industry are not aware of the idea of ESG, and the total percentage of investors that are aware of the concept of ESG is 77.10%. A statistical hypothesis test called a chi-squared test determines if two category variables have independent effects on the test statistic. In statistics, the chi-square test is used to determine the p-value. The various p-values represent the many interpretations of the hypothesis, which are listed below (Table 5):

If $p < 0.05$ in all significance tests, the two variables have a statistically significant correlation. The p-value for our chi-square finding is 0.000. This suggests that the awareness of the ESG concept in industry sectors correlates significantly. Investors use chi-square tests and crosstabs to evaluate the critical ESG concerns or elements to consider while making an investment choice (Table 6).

Table 5 Chi-square test of investors in different sectors and their awareness level regarding ESG

Chi-square tests			
	Value	df	Asymptotic significance (2-sided)
Pearson chi-square	76.093[a]	10	0
Likelihood ratio	89.715	10	0
Linear-by-linear association	0.026	1	0.871
N of valid cases	249		

[a] Nine cells (40.9%) have an expected count of less than 5. The minimum expected count is 0.92

Table 6 Sector Depiction of ESG issues

Sectors	Environmental (%)	Social (%)	Governance (%)	No consideration (%)
Real estate	38.9	27.8	33.3	–
Automobile	84.2	–	–	15.8
Information technology/ software	50	–	–	50
Pharmaceutical	49.1	–	50.9	–
Transportation	67.4	–	18.6	14
Agriculture	100	–	–	–
Hospitality	–	41.2	–	58.8
Finance/banking/insurance	61.9	–	38.1	–
FMCG	25	75	–	–
Entertainment	100	–	–	–
Education	–	–	–	100
Total	55.4	8	22.9	13.7

Table 7 Chi-square test of Industry/Sectors and ESG issues

Chi-square tests			
	Value	df	Asymptotic significance (2-sided)
Pearson chi-square	253.220[a]	30	0
Likelihood ratio	242.22	30	0
Linear-by-linear association	0.003	1	0.956
N of valid cases	249		

[a] 30 cells (68.2%) have an expected count of less than 5. The minimum expected count is 0.32

From the analysis, it can be seen that the investors who want their investment to address the ESG issues before making an investment decision—Environmental issues have the highest percentage, i.e., 55.4%, followed by Governance, i.e., 22.9%, and 13.7% of the investors does not want their investment to address the ESG issues. Investors from these sectors wish their assets to address environmental issues more than any other issue (Table 7).

A statistically significant association exists between the two variables in all significance tests if $p < 0.05$. Our chi-square result has a p-value of 0.000. This indicates a substantial correlation between the industry sectors and ESG issues.

The extent to which individual stock market investors in one of the world's rising nations are inclined toward ESG investment is demonstrated empirically by this study. This study shows that stock market investors strongly prefer environmental concerns, which contributes to the body of literature on sustainability and ESG. The findings show that investors are prepared to put money into environmentally friendly stocks, making sustainable growth possible for society, the country, and the world. The research paper consists of 11 sectors, namely Real Estate, Automobile, Information Technology, Pharmaceutical, Transportation, Agriculture, Hospitality, Finance/Banking/Insurance, FMCG, Entertainment, and Education, to study whether investors consider ESG factors while making investment decisions. The first statistical method was ANOVA, which checks investors' perceptions of ESG practices in different industries. The second statistical method used was Descriptive Analysis, under which crosstabs and chi-square tests were performed. As per the survey, investors are convinced about the long-term benefits of doing business with a company that adheres to established environmental standards, and vice versa. They have also shown an interest in considering ecological and governance concerns while making investment choices.

4 Conclusion

Globally, ESG investing is starting to become more common. This study aims to investigate the investors' ESG preferences through an initial interview and a questionnaire survey given to individual investors. The study's conclusions contribute to understanding the need for ESG concerns and will eventually spur businesses to enhance their ESG-friendly business practices. Regulators governing the stock market may base their rules and regulations on the preferences of investors. Ultimately, the country's overall ESG practices would improve with the approval and implementation of ESG rules and regulations, supporting the maintenance of social and ecological equilibrium. The investors will be guaranteed a sustainable return on investment and accountability to the environment, society, and economic equilibrium, potentially leading to sustainable development.

The good practices of businesses can provide a "Pareto improvement" for the economy as a whole. Although the study found that confident individual investors preferred ESG problems, their main reason for making this investment decision must be further investigated. Further research on this topic might shed light on the reasons behind social, functional, emotional, value-based, or revolutionary change, making it an intriguing study area. Furthermore, research is being conducted in India. It is also advised that future studies be conducted in other states and countries with different causes and concepts.

There are several restrictions on our research because of this field of study. The key drawback is that the population was limited to Indian respondents, which might substantially affect how thorough the results were overall. Moreover, the survey rating was restricted to posting on social networking sites. As a result, it reduced the possibility of hearing back from investors who were off social media. Since the most recent research is limited to individuals from a small, substantial area inside India, we could increase the study's credibility by including more people from around the globe. If retail investors from other countries are included in the research as participants and given feedback, they may significantly improve it.

Moreover, a comprehensive investigation is recommended to assess the significance of participants' economic differences in making decisions. In order to overcome the limitations of this methodology, future research may use more traditional, earlier outlets like newspapers and interviews. Further research should be conducted to assess future industry norms, trends, and young adult communication to widen the study's scope. It is also advantageous to examine how ESG aspects affect and influence an investor's decision-making process. Over the coming decades, employed young adults will be the key consumer demographic. The author calls for more study in these fields.

References

Agarwal, U. A., Datta, S., Blake-Beard, S., & Bhargava, S. (2012). Linking LMX, innovative work behaviour and turnover intentions. *Career Development International, 17*(3), 208–230. https://doi.org/10.1108/13620431211241063

Alshehhi, A., Nobanee, H., & Khare, N. (2018). The impact of sustainability practices on corporate financial performance: Literature trends and future research potential. *Sustainability, 10*(2), 494. https://doi.org/10.3390/su10020494

Bikas, E., Jurevičienė, D., Dubinskas, P., & Novickytė, L. (2013). Behavioural finance: The emergence and development trends. *Procedia – Social and Behavioral Sciences, 82*, 870–876. https://doi.org/10.1016/j.sbspro.2013.06.363

Broadstock, D., Chan, K., Cheng, L. T. W., & Wang, X. W. (2020). The role of ESG performance during times of financial crisis: Evidence from COVID-19 in China. *SSRN Electronic Journal, 38*, 101716. https://doi.org/10.2139/ssrn.3627439

Busch, T., & Hoffmann, V. H. (2011). How hot is your bottom line? Linking carbon and financial performance. *Business & Society, 50*(2), 233–265. https://doi.org/10.1177/0007650311398780

Carolina Rezende de Carvalho Ferrei, M., Amorim Sobreiro, V., Kimura, H., de Moraes, L., & Barboza, F. (2016). A systematic review of literature about finance and sustainability. *Journal of Sustainable Finance & Investment, 6*(2), 112–147. https://doi.org/10.1080/20430795.2016.1177438

Cohen, G., & Kudryavtsev, A. (2012). Investor rationality and financial decisions. *Journal of Behavioral Finance, 13*(1), 11–16. https://doi.org/10.1080/15427560.2012.653020

Cornett, M. M., Erhemjamts, O., & Tehranian, H. (2016). Greed or good deeds: An examination of the relation between corporate social responsibility and the financial performance of U.S. commercial banks around the financial crisis. *Journal of Banking & Finance, 70*, 137–159. https://doi.org/10.1016/j.jbankfin.2016.04.024

De Franco, C. (2019). ESG controversies and their impact on performance. *The Journal of Investing, 2019*(1), 106. https://doi.org/10.3905/joi.2019.1.106

de Zwaan, L., Brimble, M., & Stewart, J. (2013). Member perceptions of superannuation funds and ESG investing. *SSRN Electronic Journal*. https://doi.org/10.2139/ssrn.2316010

Escrig-Olmedo, E., Fernández-Izquierdo, M., Ferrero-Ferrero, I., Rivera-Lirio, J., & Muñoz-Torres, M. (2019). Rating the raters: Evaluating how ESG rating agencies integrate sustainability principles. *Sustainability, 11*(3), 915. https://doi.org/10.3390/su11030915

Fisher, K. L., & Statman, M. (1999). A behavioral framework for time diversification. *Financial Analysts Journal, 55*(3), 88–97. https://doi.org/10.2469/faj.v55.n3.2275

Friede, G., Busch, T., & Bassen, A. (2015). ESG and financial performance: Aggregated evidence from more than 2000 empirical studies. *Journal of Sustainable Finance & Investment, 5*(4), 210–233. https://doi.org/10.1080/20430795.2015.1118917

Hoepner, A. G. F., Oikonomou, I., Sautner, Z., Starks, L. T., & Zhou, X. (2016). ESG shareholder engagement and downside risk. *SSRN Electronic Journal*. https://doi.org/10.2139/ssrn.2874252

Ilhan, E., Sautner, Z., & Vilkov, G. (2018). Carbon tail risk. *SSRN Electronic Journal*. https://doi.org/10.2139/ssrn.3204420

Jansson, M., Biel, A., Andersson, M., & Gärling, T. (2011). Investment style and perceived drivers of adoption of socially responsible investment among Swedish institutional investors. *The Journal of Investing, 20*(3), 118–123. https://doi.org/10.3905/joi.2011.20.3.118

Jo, H., & Na, H. (2012). Does CSR reduce firm risk? Evidence from controversial industry sectors. *Journal of Business Ethics, 110*(4), 441–456. https://doi.org/10.1007/s10551-012-1492-2

Khin, E. W. S., Sedigheh, M., Teng, L. S., & Muhamad, R. (2018). Malaysia's SMEs credit industry: CSR taxanomy activities. *Audit Financiar, 16*(150), 278–284. https://doi.org/10.20869/auditf/2018/150/013

Kim, W., & Park, J. (2017). Examining structural relationships between work engagement, organizational procedural justice, knowledge sharing, and innovative work behavior for sustainable organizations. *Sustainability, 9*(2), 205. https://doi.org/10.3390/su9020205

Lee, D. D., Faff, R. W., & Rekker, S. A. C. (2013). Do high and low-ranked sustainability stocks perform differently? *International Journal of Accounting & Information Management, 21*(2), 116–132. https://doi.org/10.1108/18347641311312267

Man-Kit, B. (2015). Sustainable development and financial markets: Old paths and new avenues. *CFA Digest, 45*(9). https://doi.org/10.2469/dig.v45.n9.7

Miralles-Quirós, M. M., Miralles-Quirós, J. L., & Valente Gonçalves, L. M. (2018). The value relevance of environmental, social, and governance performance: The Brazilian case. *Sustainability, 10*(3), 574. https://doi.org/10.3390/su10030574

Nam, V. H., Luu, H. N., Nguyen, M. N., & Nguyen, D. A. (2020). The impact of corruption on the performance of newly established enterprises: Empirical evidence from a transition economy. *Borsa Istanbul Review, 20*(4), 383–395. https://doi.org/10.1016/j.bir.2020.05.006

Park, K. O. (2020). How CSV and CSR affect organizational performance: A productive behavior perspective. *International Journal of Environmental Research and Public Health, 17*(7), 2556. https://doi.org/10.3390/ijerph17072556

Pavláková Dočekalová, M., & Kocmanová, A. (2018). Comparison of sustainable environmental, social, and corporate governance value added models for investors decision making. *Sustainability, 10*(3), 649. https://doi.org/10.3390/su10030649

Pérez-Gladish, B., Benson, K., & Faff, R. (2012). Profiling socially responsible investors: Australian evidence. *Australian Journal of Management, 37*(2), 189–209. https://doi.org/10.1177/0312896211429158

Schreck, P. (2011). Reviewing the business case for corporate social responsibility: New evidence and analysis. *Journal of Business Ethics, 103*(2), 167–188. https://doi.org/10.1007/s10551-011-0867-0

Song, C., & Han, S. H. (2015). Stock market reaction to corporate crime: Evidence from South Korea. *Journal of Business Ethics, 143*(2), 323–351. https://doi.org/10.1007/s10551-015-2717-y

Sreekumar Nair, A., & Ladha, R. (2014). Determinants of non-economic investment goals among Indian investors. *Corporate Governance, 14*(5), 714–727. https://doi.org/10.1108/cg-09-2014-0102

Sultana, S., Zulkifli, N., & Zainal, D. (2018). Environmental, social and governance (ESG) and investment decision in Bangladesh. *Sustainability, 10*(6), 1831. https://doi.org/10.3390/su10061831

Thakur, B. P. S., Kannadhasan, M., Charan, P., & Gupta, C. P. (2019). Corruption and firm value: Evidence from emerging market economies. *Emerging Markets Finance and Trade, 57*, 1–16. https://doi.org/10.1080/1540496x.2019.1613643

Veld-Merkoulova, Y. V. (2011). Investment horizon and portfolio choice of private investors. *International Review of Financial Analysis, 20*(2), 68–75. https://doi.org/10.1016/j.irfa.2011.02.005

Verga Matos, P., Barros, V., & Miranda Sarmento, J. (2020). Does ESG affect the stability of dividend policies in Europe? *Sustainability, 12*(21), 8804. https://doi.org/10.3390/su12218804

von Wallis, M., & Klein, C. (2014). Ethical requirement and financial interest: A literature review on socially responsible investing. *Business Research, 8*(1), 61–98. https://doi.org/10.1007/s40685-014-0015-7

Stakeholder Engagement Strategies and Sustainable Change: A Study of Indian Companies

P. Radhika and Jyothi Kumar

Abstract "Sustainability is the pathway to prosperity for humanity and businesses alike. It offers the best chance to adjust our course."—Paul Polman. Sustainable business practices have become crucial for Indian companies aiming to balance profitability with social and environmental responsibility. These practices involve integrating sustainability into core business strategies, ensuring that operations are environmentally friendly, socially equitable, and economically viable. The dynamic business environment in India necessitates robust stakeholder engagement strategies for sustainable organizational success. Indian companies, operating within a complex socio-economic landscape, must prioritize effective stakeholder communication and collaboration to achieve long-term sustainability. This study explores various stakeholder engagement strategies implemented by Indian companies, using an assessment instrument called "Management and Organizational Sustainability Tool (MOST)," examining their impact on fostering sustainable change. By analyzing case studies and best practices, the study aims to highlight how Indian businesses can integrate stakeholder perspectives into their strategic planning, thereby enhancing resilience, trust, and overall performance in an increasingly interconnected world.

1 Introduction

Sustainable change involves long-lasting transformation that meets the needs of the present without compromising the ability of future generations. It encompasses various aspects, including environmental sustainability, social equity, and economic viability. Environmental concerns such as protecting biodiversity, reducing waste, and conserving natural resources are the focus under environmental sustainability. Addressing injustice in society and ensuring fair treatment to all citizens fall under

P. Radhika (✉) · J. Kumar
Christ University, Bengaluru, India
e-mail: radhika.p@res.christuniversity.in; jyothi.s@christuniversity.in

A. Sharma et al. (eds.), *The Synergistic Effect of Sustainable Business Practices on Corporate Performance*, https://doi.org/10.1007/978-981-96-3375-3_5

the societal aspect of sustainability. Creating economic systems that are resilient, inclusive, and capable of supporting long-term growth without depleting resources is the key focus of economic viability. All these are possible to achieve through collaboration. Therefore, "Sustainability" is a "Combined effort." It is impossible to achieve without collaborative efforts. The collaboration highly comes from the parties called "Stakeholders."

Stakeholders are a collection of individuals that are concerned about how businesses operate. They are also those who have the right to know the decisions of the company despite their diversity (White, n.d.). Stakeholders are divided into internal and external groups depending on whether they are directly or indirectly related to the business. Internal stakeholders include management, staff, and owners, while external stakeholders include the government, NGOs, suppliers, communities, customers, and the environment. Through routine meetings, questionnaires, field trips, and other activities, the expectations of the stakeholders can be monitored.

The term "stakeholder engagement" is used to describe this approach. Stakeholder engagement entails not only learning about their expectations but also considering their perspectives when making decisions. Through stakeholder involvement, sustainable performance can be accomplished. Companies must develop stakeholder engagement strategies and communicate with the stakeholders their potential for involvement. According to Manetti and Toccafondi (2012), stakeholder engagement techniques create a shared commitment between stakeholders and businesses to address societal and environmental problems. Effective stakeholder engagement acts as an essential element in achieving sustainable performance and fostering organizational innovation (Adams & McNicholas, 2007; Ayuso et al., 2011).

While the concept of stakeholder engagement is widely supported, translating it into reality poses numerous challenges. These obstacles stem from various sources, including resistance from entrenched interests, financial constraints, knowledge gaps, and the complexity of coordinating efforts among diverse stakeholders. Understanding these challenges is crucial for developing effective strategies to overcome them and ensuring that sustainability initiatives can be successfully implemented and maintained.

Indian companies have embraced sustainable business practices over the past two decades. This shift is driven by regulatory requirements, awareness about the long-term benefits of sustainability, and stakeholder expectations. Many listed companies disclose aspects related to their sustainability initiatives in standalone sustainability reports and Corporate Social Responsibility (CSR) reports. These reports are exclusively prepared for stakeholders to create a good image of the company (Gao & Zhang, 2006). The pressure for preparing a sustainability report along with financial reports generally comes from the shareholders and other stakeholders (Jørgensen et al., 2021). According to Ferrero-Ferrero et al. (2018), the highest pressure for sustainability reporting comes from employees rather than investors and environmentalists. Apart from disclosing their contributions to the environment, society, and economy, most companies have a separate section called "Stakeholder engagement." This section in the report elaborates on how the company identifies its stakeholders, understands stakeholder perspectives, develops a communication plan, fosters collaboration, and incorporates feedback given by them.

One of the main goals of sustainability reporting is to nourish stakeholder management efforts (Ortar, n.d.). Stakeholder engagement is a fundamental element in sustainability reporting since it allows companies to recognize and understand the reasonable expectations of their stakeholders (Ferrero-Ferrero et al., 2018). Sustainability reporting can have different motives and strategic choices that mainly focus on various ways to engage stakeholders. These reports also serve as a baseline for stakeholders to evaluate the business's overall performance. Maintaining good relationships with stakeholders allows companies to understand their level of dependency, which helps in deciding strategies to engage them (Herremans et al., 2016). Through stakeholder engagement, companies can aim at long-term and short-term value creation (Erin et al., 2021). Including stakeholder engagement in sustainability reports enables companies to design transparent, accountable, and responsive reports to address stakeholder issues (Kaur & Lodhia, 2018).

Creation of value by the company should be followed by the distribution of this value to the stakeholders (Aras & Crowther, 2008). Steurer et al. (2005) claim that Stakeholder Relationship Management (SRM) should be viewed as a mechanism that transmits societal issues to businesses and reduces stakeholder pressure. Involving stakeholders in decision-making can bring about better ways to tackle issues. Korsunova et al. (2016) assert that stakeholders play different roles in making sustainability-based innovation possible. There are several models developed by researchers that can embed stakeholder engagement strategies in sustainability reports. Greenwood (2007) developed a model for the stakeholder engagement process that includes different levels of adoption of stakeholder engagement by companies, such as comprehensive, excessive, little, or no engagement. Bonsón and Ratkai (2013) developed a matrix that measures the stakeholder engagement process through media presence. Rhodes et al. (2014) developed a framework for stakeholder engagement in sustainable development, claiming stakeholder engagement as a mediator between internal and external factors of the company.

Ardiana (2021) researched how stakeholder engagement strategies could be embedded with sustainability reporting practices. The existing practice of reporting shows a low level of embeddedness. There are several challenges in adopting stakeholder engagement strategies, as listed by Kaur and Lodhia (2018), including a lack of commitment among stakeholders, unwillingness to engage, and political factors. These factors affect the quality of sustainability reports. Overall, stakeholder engagement is a critical component of achieving sustainability. By involving diverse groups with varying perspectives and interests, organizations and communities can foster a more inclusive, resilient, and effective approach to sustainable development. Effective stakeholder engagement ensures that the voices of those affected by decisions are heard, leading to more equitable and accepted outcomes.

This chapter has totally six sections. The first section being the introduction tries to cover the meaning and importance of all the titles of the book chapter, Sect. 2 explains effective strategies for engaging stakeholders, Sect. 3 deals with the methodology used for conducting the research on stakeholder engagement strategies of

Indian companies, Sect. 5 highlights the important findings from the analysis, and Sect. 6 contains conclusion and future directions on the same.

2 Effective Strategies for Engaging Stakeholders

Companies have several stakeholders, both internal and external. Some of the important ones it must focus on are employees, customers, suppliers, shareholders, government, communities, and NGOs. Each of these stakeholders express varied interests in the company's activities and have a diverse degree of influence on its operations. Engaging with stakeholders is a challenging task that requires significant time and effort from companies. To effectively engage stakeholders, companies employ a variety of strategies, which differs from one to another.

Regular communications, which aims at maintaining open lines of communication through newsletters, emails, and social media updates ensures stakeholders are consistently informed about company activities and developments. This proactive approach helps build transparency and trust, keeping stakeholders engaged and aligned with the company's goals. Regular updates can cover a variety of topics such as new projects, company milestones, changes in policies, and industry news, providing a comprehensive view of the organization's progress and strategic direction.

The next best strategy would be organizing regular meetings, such as Annual General Meetings (AGMs), town halls, and focus groups, allowing direct interaction with stakeholders. These meetings provide a platform for stakeholders to voice their concerns, share feedback, and ask questions. They also enable the company to present its achievements, discuss future plans, and address any issues raised by stakeholders. This face-to-face engagement fosters a sense of inclusion and ensures that stakeholder perspectives are considered in decision-making processes.

Conducting surveys and questionnaires is a systematic way to gather stakeholder opinions and insights on various business aspects. This data-driven approach helps the company understand stakeholder needs, preferences, and satisfaction levels. It enables the companies to make informed decisions, identifies areas for improvement, and measures the impact of its strategies and initiatives. Regularly seeking feedback through these tools demonstrates a commitment to continuous improvement and stakeholder engagement.

Hosting workshops and collaborative sessions encourages stakeholders to contribute ideas and solutions to specific challenges and opportunities. These interactive forums facilitate creative problem-solving and innovation, leveraging the diverse perspectives and expertise of stakeholders for achieving sustainability goals. Collaborative workshops can address a range of topics, from strategic planning to product development, helping to align the company's efforts with stakeholder expectations and industry trends.

Implementing robust feedback mechanisms, such as suggestion boxes and online platforms, ensures continuous stakeholder input. These channels provide a convenient and accessible way for stakeholders to share their thoughts, concerns, and suggestions at any time. By promptly addressing the feedback received, the company can demonstrate responsiveness and adaptability, improving stakeholder satisfaction and trust.

Forming partnerships and alliances with key stakeholders, including NGOs, community groups, and industry associations, allows for collaborative work on sustainability goals and initiatives. These relationships can enhance the company's social and environmental impact, extend its reach, and create synergies that benefit all parties involved. Strategic alliances can also provide access to new resources, knowledge, and networks, driving sustainable growth and innovation.

Providing transparent and comprehensive reporting on company performance, sustainability efforts, and CSR activities builds trust and demonstrates accountability. Regular reports and disclosures keep stakeholders informed about the company's achievements, challenges, and future plans. Transparency in reporting ensures that stakeholders have a clear understanding of the company's operations and its commitment to ethical and sustainable practices.

Establishing advisory panels and committees comprising diverse stakeholders provides ongoing guidance and expertise on specific issues. These groups can offer valuable insights and recommendations, helping the company navigate complex challenges and seize new opportunities. Advisory panels can focus on areas such as sustainability, ethics, and community relations, ensuring that stakeholder interests are represented in strategic decision-making.

Engaging with local communities through outreach programs, volunteer activities, and support for local initiatives helps build strong, positive relationships. Community engagement demonstrates the company's commitment to social responsibility and its role as a good corporate citizen. By actively participating in community development, the company can enhance its reputation, foster goodwill, and contribute to the well-being of the communities it serves.

Offering training and educational programs to stakeholders, including employees and suppliers, enhances their understanding of sustainability practices and business goals. These programs can improve skills, knowledge, and awareness, supporting the overall effectiveness and efficiency of the organization. By investing in stakeholder development, the company can build a more capable and motivated workforce, strengthen partnerships, and promote a culture of continuous learning and improvement.

Implementing these strategies fosters stronger and more resilient relationships with stakeholders. Companies can tailor their approach by developing a "Model for Stakeholder Engagement" that best suits their unique context and objectives.

3 Model for Stakeholder Engagement

There are several models developed by academicians (Greenwood, 2007; Bonsón & Ratkai, 2013; Rhodes et al., 2014) and many industry leaders, which are successfully adopted by companies to bring in sustainable change. The researchers have used "MOST" as a scale to measure and score the stakeholder engagement strategies adopted by those companies. "MOST" was developed by "Management Sciences for Health," which is a structured and participatory process that allows companies to measure their sustainability approach. There are 5 components in MOST such as Mission, Values, Strategy, Structure, and Systems, as shown in Fig. 1 and each of them has sub-components which makes 19 parameters in Total. Each parameter is test on a scale of "1" to "4," "1" being the least and "4" being the highest. Based on the description provided in the instrument, at each level the stakeholder engagement is tested.

Organizations use the MOST instrument to evaluate these components, identify development stages, and create actionable plans for improvement, ensuring robust governance and adaptive management practices in dynamic environments.

Mission: It provides the organization's purpose and guides all levels of decision-making and strategy.

Values: These are the core beliefs and principles driving the organization's mission and staff commitment.

Fig. 1 Five components in MOST such as Mission, Values, Strategy, Structure, and Systems. Source: Author derived

Strategy: Broad approaches to define programs and activities fulfilling the mission and goals.

Structure: Organizational framework, including formal lines of authority, roles, responsibilities, and decision-making processes.

Systems: Interdependent functions like planning, human resource management, monitoring, evaluation, information management, quality assurance, financial management, revenue generation, and supply management, essential for carrying out organizational activities effectively.

MOST helps organizations assess these areas, determine their development stages, and formulate a concrete action plan for improvements by including stakeholders. This tool promotes self-assessment, strategic planning, and accountability, enabling organizations to strengthen their management systems and adapt to complex, changing conditions.

4 Research Methodology

4.1 Research Design

The study adopts descriptive research design and qualitative methods. For every parameter, we will measure the company strategies based on their engagement level with stakeholders. For example: if we are measuring the mission statement of a company, the scoring for a company will be "1" if the mission statement is unknown to stakeholders and it will be "4" if it reflects the needs of the stakeholders. Another example would be Strategy 2: Links to clients and community, where the score will be "1" if the strategies are developed without linking it to the needs of the stakeholders and "4" will be awarded if they are developed along with the participation of stakeholders. Likewise, on a scale of "1 to 4" the companies are evaluated.

4.2 Sample

The companies chosen for this study comprise the first 50 from the Nifty-fifty stock of National Stock exchange (NSE), which contains companies related to 13 different sectors of the economy. Moreover, sustainability reports are mostly published by the listed companies for their regulatory requirements.

4.3 Data Collection and Analysis

The qualitative data is extracted from the sustainability and CSR reports. The disclosures related to stakeholders are considered while searching for strategies. As explained in the research design, each company is assessed on the 19 components of "MOST" and scores are disclosed in Sect. 5 "Analysis and Results."

5 Analysis and Results

This section of the chapter deals with the results of the assessment conducted on the sample Nifty-fifty companies for the financial year 2022–2023. Table 1 exhibits the scoring technique of the first company in the list of Nifty-fifty companies, i.e. Adani Enterprises Ltd. On similar grounds, the scores were calculated for the remaining 49 companies in the Nifty-fifty list.

Most companies score highly on their mission statements, indicating a strong alignment of stakeholders with their strategic objectives. Values scores are consistently high, reflecting a solid foundation of core principles across the companies that are developed through consultation of stakeholders. In case of Strategy 1—Links to mission and values, companies like Infosys, Coal India, and Tata Steel have high scores across these dimensions, suggesting a robust approach to sustainability through stakeholder engagement. Some companies such as Bajaj Auto, Eicher Motors, and Dr. Reddy's Laboratories have lower scores, indicating potential areas for improvement in sustainability practices.

This assessment reveals that 24 out of 50 companies, as represented in Chart 1, have scored "4" on stakeholder engagement strategies. This indicates that close to 50% of the sample companies are highly involved in engaging stakeholders in their regular business conduct. This high level of engagement underscores the importance these companies place on stakeholder inclusion in their sustainability efforts, demonstrating a strong commitment to integrating stakeholder perspectives into their business operations.

Chart 2 reveals that 10% of companies scored 2, 50% scored 3, and 40% scored 4 in Strategy 3: Links to potential clients. This indicates a strong trend toward stakeholder inclusion in sustainability efforts. With 90% of companies scoring 3 or higher, it is evident that most are actively engaging potential clients in their sustainability initiatives. The fact that 40% achieved the highest score of 4 underscores a significant commitment to integrating potential clients into their strategies, fostering better business relationships and aligning with broader sustainability goals. However, the 10% scoring 2 suggests there is still room for improvement in stakeholder engagement strategies. Overall, these findings highlight that most companies are effectively including potential stakeholders in their sustainability transformations, demonstrating a proactive approach to sustainable business practices. From among the sample companies, Infosys Ltd., Tata steel Ltd., Coal India Ltd., and

Table 1 The scoring technique of the first company in the list of Nifty-fifty companies, i.e. Adani Enterprises Ltd.

Management and organization sustainability tool	AEL	
Mission	Score	Evidence
Existence and knowledge	4	The mission statement accurately reflects the current purpose and the needs of the targeted clients.
Values		
Existence and application	4	In accordance with the SEBI listing regulations, the company has developed and implemented a code of business conduct and ethics for all its board members and senior management personnel.
Strategy		
Links to mission and values	4	Corporate responsibility committee has been setup to review and revise the strategies related to ESG.
Links to clients and community	3	Stakeholder engagement ensures the needs of the customers and communities are identified and addressed.
Links to potential clients	4	The company has identified the potential stakeholders and their potential impact on business decisions.
Structure		
Lines of authority and accountability	4	Authority and responsibility are embedded with the company's culture
Governance (board of directors)	4	The board has independent and non-executive directors. The functions are effectively carried out.
Role and responsibilities	4	The role of each board member is assigned by the board committee.
Decision-making	3	Operating level decisions are taken by the chief officers for their respective departments.
Systems		
Planning	4	The company has long-term plans supporting their strategies
Communication	4	A two-way communication is ensured with all stakeholders
Human resource management	4	HR systems are in alignment with the employees
Monitoring and evaluation	3	Board performance is evaluated every year
Information management: Data collection	4	The data collected is subjected to external assurance
Information management: Use of information	3	Managers are expected to give regular feedback to the employees
Quality assurance	3	Some of the activities of the company are quality assured
Financial management	4	The company executes all its projects with annual plan and budgets
Revenue generation	4	There are long-term review-generating activities which the company can rely on
Supply management	4	The company has a supply chain that ensures efficiency

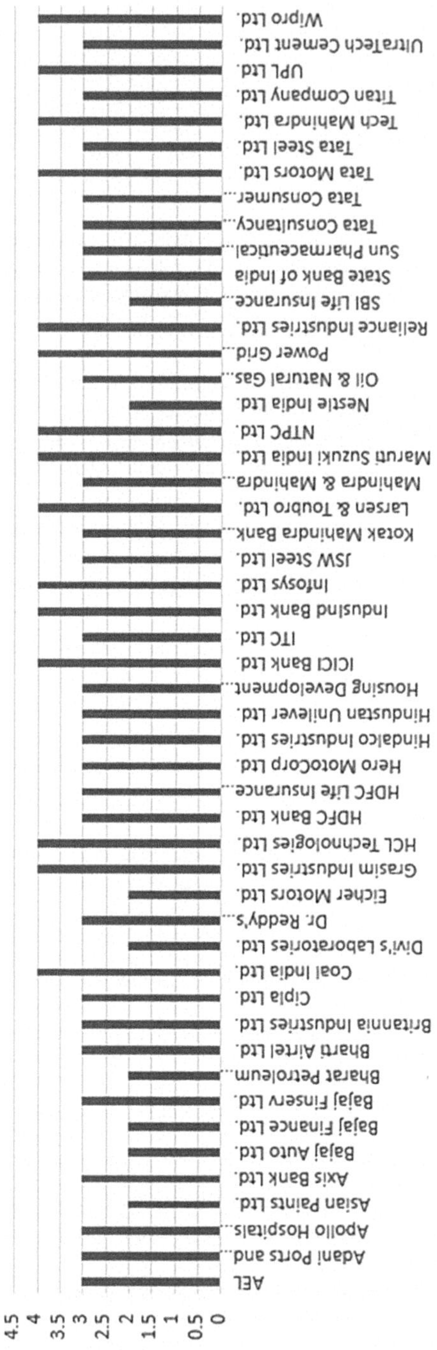

Chart 1 Strategy 2: Links to clients and community. Source: Author derived

Chart 2 Strategy 3: Links to potential clients. Source: Author derived

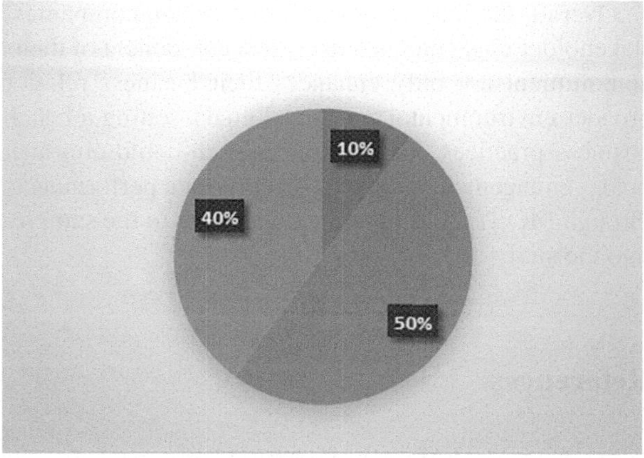

Reliance Industries Ltd. scored high on all the parameters, whereas Baja Auto and Nestle India Ltd. have lower scores.

In the case of structure, the fourth category of assessment, high scores are found in Infosys, HCL Technologies, and Reliance Industries, which suggests an effective structure of stakeholder engagement in sustainability initiatives. Lower scores in companies like Nestle India and Titan Company suggest ineffectiveness. Lastly, in Systems (including all the 10 parameters), consistently high scores are found in Infosys and Tata Steel, which indicate well-established systems and processes to support sustainability. Variability in scores for some companies like Bajaj Finance and Divi's Laboratories points to uneven implementation of systems.

6 Conclusion and Future Research

Sustainable change is a long-run process. As transformations take a long time, only business cannot achieve them. It requires persistent and collective efforts. These efforts can be pulled through stakeholders. The analysis of stakeholder engagement strategies among Indian companies reveals significant insights into the role stakeholders play in driving sustainable change. Nearly half of the sampled companies have achieved the highest scores in engaging stakeholders, indicating a robust commitment to incorporating stakeholder perspectives into their sustainability initiatives. This proactive approach is reflected across diverse industries, highlighting that sustainable practices are becoming integral to business strategies in sectors ranging from technology to energy and financial services.

Moreover, the distribution of scores in Strategy 3, where 90% of companies scored 3 or higher, underscores the growing recognition of the importance of potential client engagement in achieving sustainability goals. This trend signifies a shift toward more inclusive and comprehensive sustainability efforts, where the voices of all relevant stakeholders are considered.

Overall, the findings suggest that Indian companies are increasingly embracing stakeholder engagement as a critical component of their sustainability agendas. This commitment not only enhances their business relationships but also aligns with broader environmental and social goals, setting a benchmark for future sustainable business practices. The future research could aim at finding the impact of stakeholder engagement strategies on corporate performance, using the scores developed through MOST. Adding more companies to the sample may lead to deeper insights into the study.

References

Adams, C. A., & McNicholas, P. (2007). Making a difference: Sustainability reporting, accountability and organisational change. *Accounting, Auditing and Accountability Journal, 20*(3), 382–402. https://doi.org/10.1108/09513570710748553

Aras, G., & Crowther, D. (2008). Governance and sustainability: An investigation into the relationship between corporate governance and corporate sustainability. *Management Decision, 46*(3), 433–448. https://doi.org/10.1108/00251740810863870

Ardiana, P. A. (2021). Stakeholder engagement in sustainability reporting by fortune global 500 companies: A call for embeddedness. *Meditari Accountancy Research, 31*(2), 344–365. https://doi.org/10.1108/MEDAR-12-2019-0666

Ayuso, S., Rodríguez, M. Á., García-Castro, R., & Ariño, M. Á. (2011). Does stakeholder engagement promote sustainable innovation orientation? *Industrial Management and Data Systems, 111*(9), 1399–1417. https://doi.org/10.1108/02635571111182764

Bonsón, E., & Ratkai, M. (2013). A set of metrics to assess stakeholder engagement and social legitimacy on a corporate Facebook page. *Online Information Review, 37*(5), 787–803. https://doi.org/10.1108/OIR-03-2012-0054

Erin, O., Adegboye, A., & Bamigboye, O. A. (2021). Corporate governance and sustainability reporting quality: Evidence from Nigeria. *Sustainability Accounting, Management and Policy Journal, 13*(3), 680–707. https://doi.org/10.1108/SAMPJ-06-2020-0185

Ferrero-Ferrero, I., Fernández-Izquierdo, M. Á., Muñoz-Torres, M. J., & Bellés-Colomer, L. (2018). Stakeholder engagement in sustainability reporting in higher education: An analysis of key internal stakeholders' expectations. *International Journal of Sustainability in Higher Education, 19*(2), 313–336. https://doi.org/10.1108/IJSHE-06-2016-0116

Gao, S. S., & Zhang, J. J. (2006). Stakeholder engagement, social auditing and corporate sustainability. *Business Process Management Journal, 12*(6), 722–740. https://doi.org/10.1108/14637150610710891

Greenwood, M. (2007). Stakeholder engagement: Beyond the myth of corporate responsibility. Source. *Journal of Business Ethics, 74*(4), 315–327. https://doi.org/10.1007/sl0551-007-9509-y

Herremans, I. M., Nazari, J. A., & Mahmoudian, F. (2016). Stakeholder relationships, engagement, and sustainability reporting. *Journal of Business Ethics, 138*(3), 417–435.

Jørgensen, S., Mjøs, A., & Pedersen, L. J. T. (2021). Sustainability reporting and approaches to materiality: Tensions and potential resolutions. *Sustainability Accounting, Management and Policy Journal, 13*(2), 341–361. https://doi.org/10.1108/SAMPJ-01-2021-0009

Kaur, A., & Lodhia, S. (2018). Stakeholder engagement in sustainability accounting and reporting: A study of Australian local councils. *Accounting, Auditing and Accountability Journal, 31*(1), 338–368. https://doi.org/10.1108/AAAJ-12-2014-1901

Korsunova, A., Goodman, J. C., & Halme, M. (2016). Activities and roles of stakeholders in sustainability-oriented innovation of firms. *Academy of Management Proceedings, 2016*(1), 15594. https://doi.org/10.5465/ambpp.2016.15594abstract

Manetti, G., & Toccafondi, S. (2012). The role of stakeholders in sustainability reporting assurance. Source. *Journal of Business Ethics, 107*(3), 363–377. https://doi.org/10.1007/sl0551-01. https://msh.org/resources/management-and-organizational-sustainability-tool-most/

Ortar, L. (n.d.). Materiality matrixes in sustainability reporting: An empirical examination. *Journal of Strategic Innovation and Sustainability, 15*(1).

Rhodes, J., Bergstrom, B., Lok, P., & Cheng, V. (2014). A framework for stakeholder engagement and sustainable development in MNCs. *Journal of Global Responsibility, 5*(1), 82–103. https://doi.org/10.1108/JGR-02-2014-0004

Steurer, R., Langer, M. E., Konrad, A., & Martinuzzi, A. (2005). Corporations, stakeholders and sustainable development I: A theoretical exploration of business-society relations. *Journal of Business Ethics, 61*(3), 263–281. https://www.jstor.org/stable/25123621

White, A. L. (n.d.). Why we need global standards for corporate disclosure. https://www.jstor.org/stable/27592143

Consumers' Perspective Toward Fintech Disruption in Financial Services—A Structural Equation Modeling Approach

T. Mahesh Kumar, K. Abinaya, Manjari Sharma, and Sharad Gupta

Abstract For nearly a decade, banks have faced an unprecedented array of challenges, with shrinking revenues and intense pressure from consumers and regulators. Compounding this tough environment is a powerful new force that has emerged to challenge banks, financial technology (FinTech) firms. These innovative start-ups are striving to take a share of financial services' customers, products, and revenue. "Financial Consumer Demands for Tomorrow's Digital Bank," investigated what consumers' value from their main financial service provider and how well banks are meeting consumers' needs. The study revealed a demand for value-added services that supplement traditional banking products, while also solidifying the importance of seamless, anytime/anywhere, omni-channel service channels. This study is such an attempt to analyze the consumer perspective based on the Fintech Disruption in Financial Services in Pune city among 150 respondents. The tools used for this study are ANOVA, regression analysis, and structural equation modeling (SEM). Hence, this study concludes that financial institutions have a lot on their plate: emerging competitors, shifting demographics, rising customer expectations, and changing regulations. Technology offers solutions, allowing financial institutions to cut costs and become more efficient at what they do.

T. Mahesh Kumar
Department of Commerce, Christ (Deemed to be University), Pune, India
e-mail: maheshkumart@christuniversity.in

K. Abinaya
Department of Commerce (CA), KPR College of Arts Science & Research, Coimbatore, India
e-mail: abinaya.k@kprcas.ac.in

M. Sharma (✉)
School of Commerce, Finance and Accountancy, Christ University, Pune, India
e-mail: manjari.sharma@christuniversity.in

S. Gupta
School of Business and Management, Christ (Deemed to be University), Pune, India
e-mail: sharad.gupta@christuniversity.in

A. Sharma et al. (eds.), *The Synergistic Effect of Sustainable Business Practices on Corporate Performance*, https://doi.org/10.1007/978-981-96-3375-3_6

1 Introduction

The financial services industry has seen drastic technology-led changes over the past few years. Many executives look to their IT departments to improve efficiency and facilitate game-changing innovation while somehow also lowering costs and continuing to support legacy systems. Meanwhile, FinTech start-ups are encroaching upon established markets, leading with customer friendly solutions developed from the ground up and unencumbered by legacy systems. Customers have had their expectations set by other industries; they are now demanding better services, seamless experiences regardless of channel, and more value for their money. Regulators demand more from the industry too and have started to adopt new technologies that will revolutionize their ability to collect and analyze information. And the pace of change shows no signs of slowing. In our latest Global CEO Survey, across all sectors business leaders told us the speed of technological change is one of their biggest concerns.

Ishtiaq Ahmad Bajwa et al. (2022) conducted a meta-literature review examining the past, present, and possible future trends of Fintech research using 360 selected articles published between 2006 and June 2020. Both quantitative and qualitative techniques were applied. In the quantitative approach, a bibliometric citation analysis using HistCite and VOSviewer software was conducted and the qualitative analysis covered the identification of four main research streams related to (i) Bitcoin and digital currency, (ii) crowdfunding, (iii) mobile payment, and (iv) blockchain. In fact, in financial services, 70% of the leaders told us the speed of change in technology was a concern. One factor is that the time it takes to go from breakthrough technology to mass-market application is collapsing. For example, in the United States, it took the telephone 76 years to be adopted by half the population. By contrast, the smartphone did it in less than 10 years. We are now watching blockchain move from a notebook sketch to an established technology in a tiny fraction of the time it took for the Internet to be accepted as a standard tool. Indeed, technology-driven change is so pervasive that no financial institution is immune. There are many large forces sweeping society, from demographic and social changes to shifts in global economic power. But one force in particular namely, technological breakthroughs having a disproportionate effect on financial services.

2 Review of Literature

Christensen et al. (2015) clarified some of the misconceptions about disruption. Disruption is a process that takes time: complete substitution, if it happens at all, may take decades. This dynamic helps to explain why incumbents, who have the resources to defend their established positions, often overlook disruptors, who gradually erode incumbents' margins and ultimately profits. Disruptors often build business models that are quite different from those of incumbents. Success in the marketplace is not an inherent feature of disruptive innovation theory: some disruptive innovations will succeed, while others will fail. **Chuen et al.** (2015) discussed the LASIC (Low margin, Asset light, Scalable, Innovative, and Compliance easy) principles for successful fintech firms, such as Alibaba and M-PESA, and examine the benefits of investing in fintech start-ups for financial inclusion. Academic research that focuses on an industry-level analysis of the extent of disruption that fintech innovations can bring to the financial services system is still sparse. **According to PwC** (2016), fintech is "the segment that is at the intersection of the financial services and technology sectors where technology-focused start-ups and new market entrants innovate the products and services currently provided by the traditional financial services industry." Blockchain is a type of an open, distributed ledger technology capable of recording anonymous peer-to-peer transactions of value (e.g., money and other assets) in a verifiable, immutable way. The ledger can be programmed to trigger transactions automatically. **Gomber et al.** (2018) presented a new fintech innovation mapping approach that enables the assessment of the extent to which there are changes and transformations in four areas of financial services. They also discussed operations management in financial services and the changes occurring; technology innovations that have begun to leverage the execution and stakeholder value associated with payments, crypto currencies, block chain, and cross-border payments; multiple innovations that have affected lending and deposit services, peer-to-peer (P2P) lending, and social media use; issues with respect to investments, financial markets, trading, risk management, robo-advisory, and services influenced by blockchain and fintech innovations.

Emily Lee (2022) examined about the banks' de-risking practices inside Hong Kong's Anti-Money Laundering/Countering the Financing of Terrorism (AML/CFT) regime, a problem that has created considerable tension between the demands of AML/CFT prevention and those of financial inclusion. The authors argued that tiered account services are an important first step toward financial inclusion, culminating in the introduction of simple bank accounts by some banks to mitigate the effect of de-risking. While proposed solutions such as the know-your-client utility system and central data repository may contribute to a digital financial inclusion framework, they are not tailored to solve a specific problem (de-risking). **Valeria Stefanelli and** Manta (2023) inquired the strategic positioning of European banks toward their digital transformation strategies, by analyzing the decision-making processes that occurred between 2015 and 2022. A qualitative analysis of partnerships and the adoption of Application Programming Interfaces (APIs) development in support of new service models was carried out. Results have relevant policy implications for regulators, linked to the business evolution and the risks of outsourcing, and managerial implications for the followers, specifically on the plan of service integration to diversify and boost their activities in the segment of customer relationship management and care, providing a better user experience. **Laurie Hughe et al.** (2023) aimed to develop novel insight to the "lived in" impact of digital transformation within the FinTech industry from a factor interdependency perspective. They also adopted a mixed methods approach incorporating interpretive structural modeling, analytical hierarchy process, and interviews with expert participants, to offer a unique perspective on the challenges and unintended consequences of industry-level technological change. They highlighted the high levels of interdependency and priority for challenges related to the investment in products and infrastructure for new markets, criticality of stakeholder support, and development of a digital mindset for the adoption of new technologies. **Linda Court Salisbury et al.** (2023) proposed a novel, theory-driven definition of consumer financial vulnerability (CFV) as the risk of incurring future harm, given the consumer's current access to various financial resources. A new conceptual framework decouples "vulnerability" from "harm" to distinguish the state of CFV, its determinants (access to various interdependent financial resources), and the constructs it foreshadows (multiple interconnected forms of realized harm). It is also followed that five research propositions such as (1) financial resource volatility plays a vital role in CFV, (2) recovering from harm requires more financial resources than preventing harm, (3) a multiperiod lens is needed to assess CFV accurately, (4) greater financial resource access can increase CFV, and (5) generalized financial literacy is not a panacea for mitigating CFV. The propositions and their implications for marketing strategy, public policy, and consumer well-being offer a rich research agenda. They also proposed a measure of CFV—the probability that financial resources are insufficient to meet or exceed a harm threshold—for future empirical investigations.

3 Statement of the Problem

For a long time, new market entrants found it difficult to break into the financial services industry. They had strong compliance systems in place to manage ever-increasing regulations, and they had the client base and resources to prosper even in tough economic conditions. Well, not any more. FinTech disruptors have been finding a way in. Disruptors are fast-moving companies, often start-ups, focused on a particular innovative technology or process in everything from mobile payments to insurance. And, they have been attacking some of the most profitable elements of the financial services value chain. This has been particularly damaging to the incumbents who have historically subsidized important but less profitable service offerings. In the late 1990s, when companies began to realize the Internet's potential power, e-commerce investment and experimentation soared. And despite the "dot com crash," it is unlikely that anyone would deny just how revolutionary the technology has proved to be. Today, there are curious similarities with blockchain both in how companies are being funded and how they are exploring use cases. Hence, this study is such an attempt to analyze the consumers' perspective based on the Fintech Disruption in Financial Services.

4 Objectives of the Study

1. To examine the consumers' perspective on the fintech disruption in financial services.
2. To develop a regression model for escalating the level of digital payments with the insurance, investment management, and market provisioning.

5 Research Methodology

5.1 Collection of Data

The primary data of this study were collected from the respondents in Pune City with the help of well-structured questionnaire. This study is developed to understand the consumers' perspective based on the Fintech Disruption in Financial Services.

5.2 Sampling Method

Simple random sampling method was adopted to analyze the consumers' perspective based on the Fintech Disruption in Financial Services; for the purpose of this research work, 150 respondents were selected.

5.3 Tools Used for Analysis

The primary data have been analyzed by using the statistical tools like ANOVA, regression analysis, and structural equation modeling.

6 Analysis and Interpretation

Regression Analysis on Digital Payments and Market Provisioning
Table 1 summarizes the model summary of the impact of Awareness on Attitude. When Attitude is a dependent variable, $R = 0.810$ which means that there is a strong relationship. R-square is 0.646 indicating that 64.6% of performance variation is accounted for the combined linear impact of independent variables. Adjusted R-square value is 0.641, implying that the model has accounted for 64.1% of the

Table 1 Model summary of digital payments and market provisioning

Model summary[b]					
Model	R	R-square	Adjusted R-square	Std. error of the estimate	Durbin–Watson
1	.810[a]	0.646	0.641	0.31876	2.568

[a] Predictors: (constant), market provisioning
[b] Dependent variable: Digital payments

Table 2 ANOVA of digital payments and market provisioning

ANOVA[a]						
Model		Sum of squares	Df	Mean square	F	Sig.
1	Regression	0.266	1	0.266	0.975	.315[b]
	Residual	110.075	149	0.262		
	Total	110.341	150			

[a] Dependent variable: Digital payments
[b] Predictors: (constant), market provisioning

Table 3 Coefficients of market provisioning on digital payments

Coefficients[a]						
		Unstandardized coefficients		Standardized coefficients		
Model		B	Std. error	Beta	T	Sig.
1	(constant)	2.375	0.114		109.751	0.000
	Market provisioning	0.058	0.060	0.192	0.939	0.315

[a] Dependent variable: Attitude

variance in the criterion variable. The value of Durbin–Watson statistic is 2.568, representing that the model is suffering from auto-correlation.

$H_0 1$: There is no significant effect of digital payments and market provisioning.

Table 2 explains the ANOVA for digital payments and market provisioning. The significant value for the above model is greater than 0.05, i.e. 0.315 which considers digital payments as a dependent variable and market provisioning as an independent variable. Hence, the null hypothesis is accepted. It is concluded that there is no significant effect of digital payments on market provisioning.

Table 3 shows the coefficients of the impact of market provisioning on digital payments. It implies that market provisioning is not significant at 5% significance level. Beta value shows that market provisioning has a positive relationship with digital payments. It has positive impact on digital payments which means market provisioning causes impact on digital payments (Tables 4 and 5).

Table 4 Measurement of reflective constructs

Test		Indicator reliability	Convergent validity	Internal consistency reliability	Reliability	Discriminant validity
Construct		Factor loading	AVE	Composite reliability (R^2)	Cronbach's alpha	Sq. root of AVE
Insurance	Connected insurance	0.882	0.808	0.894	0.764	0.8988
	Insurance disaggregation	0.916				
Market provisioning	New market platforms	0.939	0.848	0.918	0.823	0.9208
	Smarter, faster machines	0.903				
Investment management	Empowered investors	0.904	0.772	0.871	0.707	0.8786
	Process externalization	0.852				

Table 5 Outer model measures of reflective constructs

	Sqrt of AVE should be > Correlations of latent variables			
Discriminant validity	Insurance	Market provisioning	Investment management	Sq.root of AVE
Insurance	1			0.8988
Market provisioning	0.808	1		0.9208
Investment management	0.772	0.848	1	0.8786

7 Outer Model Assessment for Reflective Constructs

The reflective constructs of study, namely insurance, market provisioning, and investment management are assessed (Fig. 1):

As illustrated in Table 6, bootstrapping technique is applied to obtain the t-statistic value using 500 samples (as the sample size of the study is 150) and 450 resample (thrice the sample size) as prescribed by Hair (2014).

From Table 7, it is inferred that the t-statistics for both reflective construct paths and formative construct path are significant as their t-values are greater than 1.96 at 5% level of significance. Hence, the model developed supports the theory with the statistical proof (Tables 8 and 9).

Interpretation for R^2, the Coefficient of Determination (Table 5) R^2 is a measure of the model's predictive accuracy. Here, the R^2 value in Table 5, for the endogenous constructs insurance (0.336), market provisioning (0.417), and investment management (0.300) proves that there is moderate fit of the model.

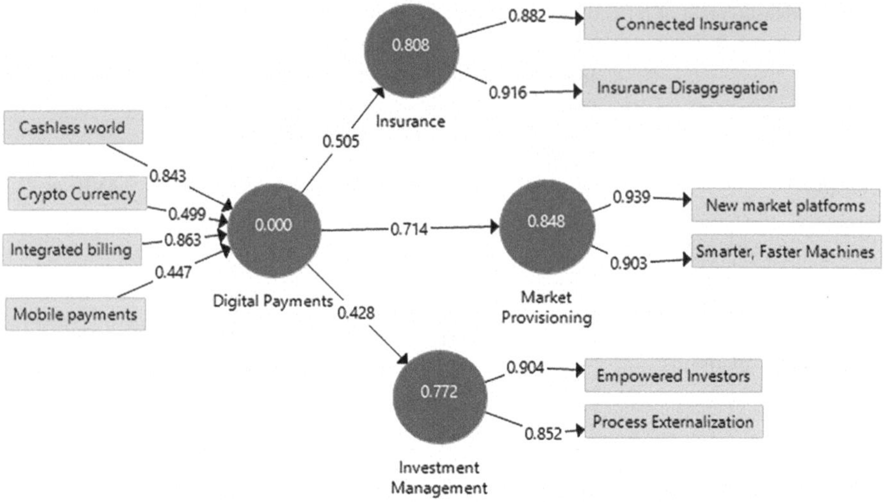

Fig. 1 Measurement model

Table 6 Measurement model for formative constructs

Construct	Indicator	Outer loading	t-stat for outer loading	Outer weight	t-stat for outer weight
Digital payments	Cashless world	0.843	15.743	0.565	4.917
	Crypto currency	0.499	4.493	0.043	2.362
	Integrated billing	0.863	17.392	0.606	7.342
	Mobile payments	0.447	3.926	0.050	1.983

Interpretation of Q^2, the Predictive Relevance (Table 5) Q^2 is not to be calculated for formative endogenous construct (Hair, 2014, 2011) (Hair, 2014). Hence, it is not calculated for digital payments (insurance, market provisioning, and investment management). Q^2 values for insurance (0.416), market provisioning (0.445), and attitude (0.396) are greater than zero and hence, it is concluded that digital payments can be predicted. This accomplishes the research objective.

Interpretation of f^2 and q^2 (Table 6) The f^2 and q^2 are calculated using Cohen's formula in Table 6. The ultimate understandable results are depicted in table. The effect values of endogenous constructs shown in the table explain that digital payments can be predicted by integrated billing, cashless world, and mobile payments with high accuracy, while by crypto currency with medium accuracy (Fig. 2).

Table 7 Measurement model for reflective and formative constructs

	Loadings (to be referred for reflective)		Weights (to be referred for constructive)	
Path	Path coefficient (original sample (0))	t-statistics	Path coefficient	t-statistics
Reflective constructs				
Connected insurance Insurance	**0.882**	**32.241**	0.511	19.933
Insurance disaggregation insurance	**0.916**	**70.204**	0.599	18.852
New Market platforms market provisioning	**0.939**	**109.751**	0.601	20.454
Smarter, faster machines market provisioning	**0.903**	**38.120**	0.482	22.975
Empowered investors investment management	**0.904**	**60.798**	0.625	13.788
Process externalization investment management	**0.852**	**22.002**	0.510	13.927
Cashless world digital payments	**0.843**	**15.743**	0.565	4.917
Crypto currency Digital payments	**0.499**	**4.493**	0.043	2.362
Integrated Billing digital payments	**0.863**	**17.392**	0.606	7.342
Mobile payments digital payments	**0.447**	**3.926**	0.050	1.983

Table 8 Calculation of R^2 effect value and Q^2 effect value

	R-square included	Q-square included
Insurance	0.336	0.416
Market provisioning	0.417	0.445
Investment management	0.300	0.396
For digital payments	**R-square excluded**	**Q-square excluded**
Digital payments—Cashless world	0.505	Q-square not to be calculated for formative constructs
Digital payments—Crypto currency	0.428	
Digital payments—Integrated billing	0.714	
Digital payments—Mobile payments	0.565	

Table 9 Effect size—Consolidated table

Dependent variable	Independent variable	f^2	f^2 effect	q^2	q^2 effect
	Insurance	0.336	Medium	0.416	Medium
	Market provisioning	0.417	Medium	0.445	Medium
	Investment management	0.300	Medium	0.396	Medium
Digital payments	Cashless world	0.505	Large	NA	NA
	Crypto currency	0.428	Medium	NA	NA
	Integrated billing	0.714	Large	NA	NA
	Mobile payments	0.565	Large	NA	NA

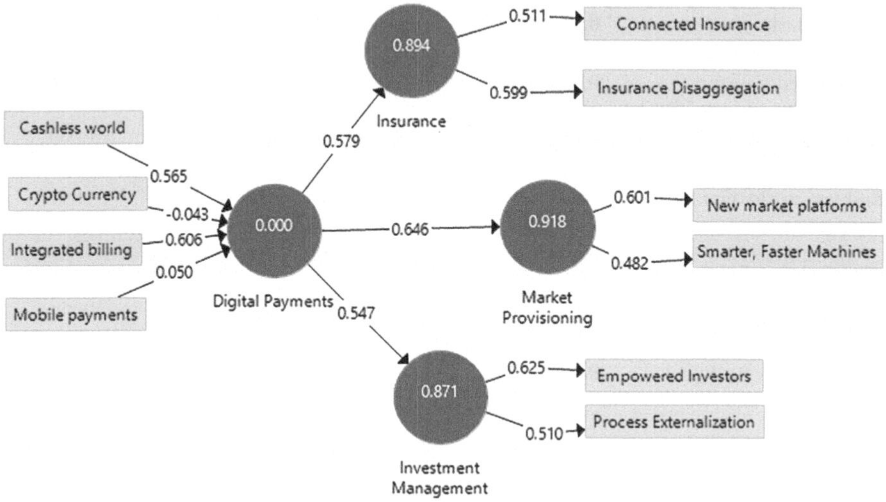

Fig. 2 Inner model

8 Interpretation

8.1 Reporting the Inner Model

- **Adjusted R-square:** The adjusted R^2 explains the coefficient of determinant and the variance in predicting the overall model (Hair 2014; Chin, 2010), suggesting 0.25 (weak), 0.50 (moderate), and 0.75 (strong) indices, respectively. In the current model, the adjusted R-square is 0.646 for digital payments, which means that insurance, market provisioning, and investment management together explain up to 65% (with higher moderate) of variance of digital payments.
- **Effect size f^2:** Effect size examines how large is the influence of exogenous variable on the endogenous variable, suggesting 0.02 (weak), 0.15 (moderate), and 0.35 (strong) indices, respectively. In the model developed, the f^2 effect size is digital payments to cashless world (0.505—moderate), digital payments to

crypto currency (0.428—medium), digital payments to integrated billing (0.714—highly moderate), and digital payments to mobile payments (0.565—moderate). Effect size for behavior (being formative construct) is not to be calculated (Hair, 2014).

- **t-statistics:** Bootstrapping is applied for the significance of the path coefficient with two tails at 5% level of significance. The t-statistics for all the path coefficients are greater than 1.96, confirming the significant path relationship with the constructs.

9 Findings

1. There is no significant effect of digital payments on market provisioning.
2. Beta value shows that market provisioning has a positive relationship with digital payments.
3. In the current SEM model, the adjusted R-square is 0.646 for digital payments, which means that insurance, market provisioning, and investment management together explain up to 65% (with higher moderate) of variance of digital payments.
4. The t-statistics for all the path coefficients are greater than 1.96, confirming the significant path relationship with the constructs.

10 Suggestions

1. The Government should establish a policy framework for formalizing alternative data sources for credit scoring.
2. The RBI should establish a mandatory MSME referral Programme to facilitate MSME lending by alternative lenders.
3. Financial service players should identify core consumer needs and build segment-specific offerings targeting lower-income households to drive Financial Service (FS) solutions at the point of consumption and improve financial inclusion.
4. Consumers should build their financial and digital literacy to improve their knowledge of Financial service products and avoid becoming victims of cybercrime.

11 Conclusion

A flagship initiative of the Government, Jan Dhan Yojana scheme aims to drive financial inclusion across the country. This has resulted in a significant uptick in the number of people with bank accounts in India. The Government's push to improve digital literacy across the country through a bouquet of initiatives spanning infrastructure, literacy, and ease of accessing digital services has had a significant impact in improving the overall digital maturity of the Indian populace. However, the road to collaboration is not free from obstacles. Differences in the work culture, business models, and security continue to pose key challenges. While incumbents struggle with the pace of innovation and the rigidity of their legacy systems, start-ups face legal, bureaucratic, and cultural issues while working with institutions. Despite these challenges, collaboration has become a highly promising avenue for growth and is expected to remain so in the future, leading to increased adoption of FinTech solutions among consumers.

References

Bajwa, I. A., Ur Rehman, S., Iqbal, A., Anwer, Z., Ashiq, M., & Khan, M. A. (2022). Past, present and future of FinTech research: A Bibliometric analysis. *SAGE Open, 12*(4), 21582440221131242. https://doi.org/10.1177/21582440221131242

Chin, W. W. (2010). A comparison of approaches for the analysis of interaction effects between latent variables using partial least squares path modeling. *SEM, 17*(1), 107–120.

Christensen, C. M., Raynor, M. E., & McDonald, R. (2015). Disruptive Innovation. *Harvard Business Review, 93*(12), 44–53.

Chuen, K., Lee, D., & Teo, E. (2015). Emergence of FinTech and the LASIC principles. *Journal of Financial Perspectives, 3*(3), 24–36.

Gomber, P., Kauffman, R., Parker, C., & Weber, B. W. (2018). On the Fintech revolution: Interpreting the forces of innovation, disruption, and transformation in financial services. *Journal of Management Information Systems, 35*, 220–265.

Hair, J. R. (2011). PLS-SEM: Indeed, a silver bullet. *Journal of Marketing Theory and Practice, 19*(2), 139–152.

Hair, J. F. M. (2014). Partial least squares, structural equation modeling (PLS-SEM): An emerging tool in business research. *European Business Review, 26*, 106–121.

Hughes, L., Seddon, J. J., & Dwivedi, Y. K. (2023). Disruptive change within financial technology: A methodological analysis of digital transformation challenges. *Journal of Information Technology, 39*(4), 8. https://doi.org/10.1177/02683962231219512

Lee, E. (2022). Technology-driven solutions to banks' de-risking practices in Hong Kong: FinTech and blockchain-based smart contracts for financial inclusion. *Common Law World Review, 51*(1–2), 83–108. https://doi.org/10.1177/14737795211071095

PwC. (2016). Financial services technology 2020 and beyond: Embracing disruption. Retrieved from https://www.pwc.com/gx/en/financial-services/assets/pdf/technology2020-and-beyond.pdf

Salisbury, L. C., Nenkov, G. Y., Blanchard, S. J., Hill, R. P., Brown, A. L., & Martin, K. D. (2023). Beyond income: Dynamic consumer financial vulnerability. *Journal of Marketing, 87*(5), 657–678. https://doi.org/10.1177/00222429221150910

Stefanelli, V., & Manta, F. (2023). Digital financial services and open banking innovation: Are banks becoming 'invisible'? *Global Business Review*. https://doi.org/10.1177/09721509231151491

A Systematic Review and Bibliometric Analysis on the Future Directions of Environmental Sustainability and Green Finance

Varun Kesavan and Aruna Polisetty

Abstract This paper aims to critically discuss the intersection between environmental sustainability and green finance through systematic review and bibliometric analysis. The study objectives are to address these gaps with the help of analysis of other sources containing bibliographic records, having considered 386 papers in the Scopus database. The results based on quantitative data show that it has been rapidly developing during the last 6 years, as more than 96% of the identified articles belong to the period from 2017 to 2024. Among all the countries, China, India, and Indonesia can be considered leaders by the number of publications in the field of environmental sustainability and green finance. The existent paper distribution across journals is that "Resources Policy" is the most influential journal in this field that contributes 20% of the papers about environmental sustainability and green finance with 77 papers followed by "Sustainability (Switzerland)" with 33 articles, "Energy Economics" with 25 articles, and finally "Journal of Cleaner Production" with 20 articles. Al-Faryan, M. A. S is the most productive author with 5 contributions to the journal while authors Ibrahim, R. L. and Lin, J. H have contributed 4 articles to the journal. This paper is a good contribution to the literature on environmental sustainability and green finance since co-occurrence word analysis shows that the important fields of interest are "green finance," "fintech," "sustainable development," "green innovation," "natural resources," "climate change," "green bonds," and "sustainable finance." This paper aims to add to the growing knowledge about environmental sustainability and specifically green finance.

V. Kesavan (✉) · A. Polisetty
VIT Business School, VIT University, Vellore, TN, India
e-mail: varun.kesavan2020@vitstudent.ac.in; arunapolisetty@vit.ac.in

A. Sharma et al. (eds.), *The Synergistic Effect of Sustainable Business Practices on Corporate Performance*, https://doi.org/10.1007/978-981-96-3375-3_7

1 Introduction

Sustainable finance has become an effective way enabler for financing the environmental sustainability to support investments for the environment (Chang et al., 2023; Sinha et al., 2021; Zakari & Khan, 2022; Peng & Zheng, 2021; Falcone & Sica, 2019). Currently, it is appreciated for the commitment toward lowering carbon footprint, promoting green economy development, and managing social and environmental challenges in the international financial milieu (Huynh et al., 2022; Koczar et al., 2020). Green bonds are among many other sustainable financing instruments being implemented in different countries for upcoming Sustainable Development Goals (SDGs) as a way of embracing environmental and social corporate responsibility (Sinha et al., 2021; Zhang et al., 2022). As a consequence, the study of green finance has gained popularity among scholars as many researchers have focused on this area in the recent past (Khan et al., 2024).

The primary purpose of this systematic review is to evaluate existing academic articles on environmental sustainability and sustainable finance by applying bibliometric analysis to identify new insights and trends in this area of research. This research will attempt to offer such a synthesis in order to map the current state of knowledge, document areas of previous research and various subtopics, and suggest potential future avenues for research (Ahmi & Nasir, 2019; Ellegaard & Wallin, 2015; Roig-Tierno et al., 2017). In a broad sense, by conducting a systematic review, scattered data can be reviewed focused and integrated knowledge, practices, or research initiatives in green finance and environmental sustainability could be brought and aligned (Khan et al., 2024). The objective is to build up a sound and consolidated knowledge and to ensure that further development of this field will find here the necessary platform for further research.

RQ1. How fast is new information, especially from scientific articles, being produced concerning environmental sustainability and sustainable finance on an annual basis?

RQ2. There is interest in identifying which authors, documents, journals, and countries are most effective in terms of writing about environmental sustainability and the literature on sustainable finance.

RQ3. Where does the current literature about environmental sustainability and sustainable finance stand and where might it be most useful to focus research?

RQ4. In what ways and according to what trends are the topics for research in the sphere of environmental sustainability and sustainable finance being developed?

2 Literature Review

This paper on green finance and its consequences in terms of the environment has elicited considerable interest in the recent past. According to a study conducted by Khan et al. (2024), they have claimed that environmental sustainability and sustainable finance are important for environmental sustainability. For instance, Wu et al. (2021) targeted cessation of weather transformation mitigation by analyzing effects of sustainable financing within G7 and E7 nations. According to their research, several critical points were identified in the analysis of climate change and the effects of environmental pollution such as the need for green finance. Based on the research done by Muganyi et al. (2021), they also talked about how effective sustainable finance strategies are in China. They found that these strategies have cut down on the emissions of manufacturing vapors and that China could become a world leader by adopting them.

Further, there exists a study by Meo and Karim (2022) to highlight the role of sustainable finance in managing CO2 emissions that have employed quantile-on-quantile regression analysis to crosswise ten economies with sustainable finance listing. Their investigation revealed diverse relationships depending on the conditions within the market places and characteristics of a certain country; therefore, financing for sustainable development is a crucial factor in reducing emissions of CO2 and the need to use it as an appropriate model for combating environmental challenges. Furthermore, utilizing the Stochastic Impact by Regression on Population, Afuence, and Technology (STIRPAT) technique, Rasoulinezhad and Taghizadeh-Hesary (2022) argued that green bonds are effective to catalyze green energy projects and decrease CO_2 emissions, and sustainable long-term policies are essential (Khan et al., 2024).

Similarly, another bold study to determine sustainable finance was done by Ren et al. (2020) which evaluated the function of sustainable finance in certifying carbon emissions through building a sustainable finance index. They also agreed that pushing for green finance's advancement gives way to increase uptake of other than fossil energy and decreased carbon intensity. Nawaz et al. (2020) looked at the relationship between sustainable finance and investment risk in emerging and developing countries. They used Diference in Diferences (DID) to test the causal effects of the forces that affect sustainable finance and climate change mitigation. They only discovered the different influence of sustainable financing on climate change mitigation; for this reason, they report that this relationship is complex (Khan et al. 2024).

Also, to be able to identify positive and negative effects of the growing focus on environmental sustainability and green finance, it is necessary to analyze them. For instance, Falcone and Sica (2019) highlighted financial challenges that might affect green companies' investment plans, thus lighting up aspects of green finance complexity. Using the context of energy consumption structure transformation in China, Gu et al. (2023) discussed the impact of green finance on that change noting that to further enhance this impact, the hurdles should be addressed while taking advantage of the opportunities arising from green finance.

Even if sustainable finance is often connected to positive environmental outcomes such as climate change and sustainable development, various works raised concerns about its efficacy. According to Chen and Chen (2021) and Fang and Shao (2022), sustainable finance has proven ineffective in lowering carbon emissions and improving the definitions of industrial pollution control events. Sustainable finance as pointed out by Li et al. (2022) promotes monetary development coupled with a reduction in the pollutive process; nonetheless, it poses a threat to clean energy development. Zakari and Khan (2022) further noted that the concept of green finance could have detrimental impacts to environmental sustainability; thus, it is about time that the negative side effects of green finance are also deliberated on.

In summary, the literature review on environmental sustainability and green finance highlights the varied parts that green finance plays. Examining all previous research on this topic is crucial as its importance grows. This will help us figure out what the trends are, what we already know, what we should study next, and how we can help this important and quickly growing area grow (Khan et al., 2024).

3 Materials and Methods

This paper intends to assess changes in the literature from the last 5 years with regard to environmental sustainability and green finance by category analysis and bibliographic analysis. Using these approaches is useful in a particular scientific field, as they aid in recognizing the existing and expanding knowledge and its weak spots. Also, bibliometric analysis standardizes the valuation of academic works in terms of quality and quantity with visions into the study tendencies, evaluating the impact of research, and recognizing new themes and concepts to be expected in future research (Ellegaard & Wallin, 2015; Roig-Tierno et al., 2017; Zyoud et al., 2018; Khan et al., 2024).

4 Results and Discussion

4.1 Source and Data Collections

This study used the Scopus database; the strategy for identification included using keywords such as environmental sustainability and green finance. Scopus was used because it is among the most valued indexes in the global academic community, which is one of the major and most reliable sources in the sphere of academic research (Schotten et al., 2017; Sharif et al., 2020; Khan et al., 2024). The specific importance lies in its wide attention to technical works across diverse emerging arenas, making it an ideal choice for revealing a comprehensive bibliometric analysis in the field. Keywords for the search queries were selected with utmost care to ascertain comprehensiveness where the search string contained plain relevant keywords related solely to green finance and environmental sustainability including the terms "green finance" and "environmental sustainability," in order to capture all facets of the subject. The systematic search yielded in total of 386 documents. In this way, all the bibliographic records that were selected for the study were examined in detail and those that were not seen as relevant to the general theme of the study were excluded. Further, Fig. 1 provides a worthy illustration of the tracked

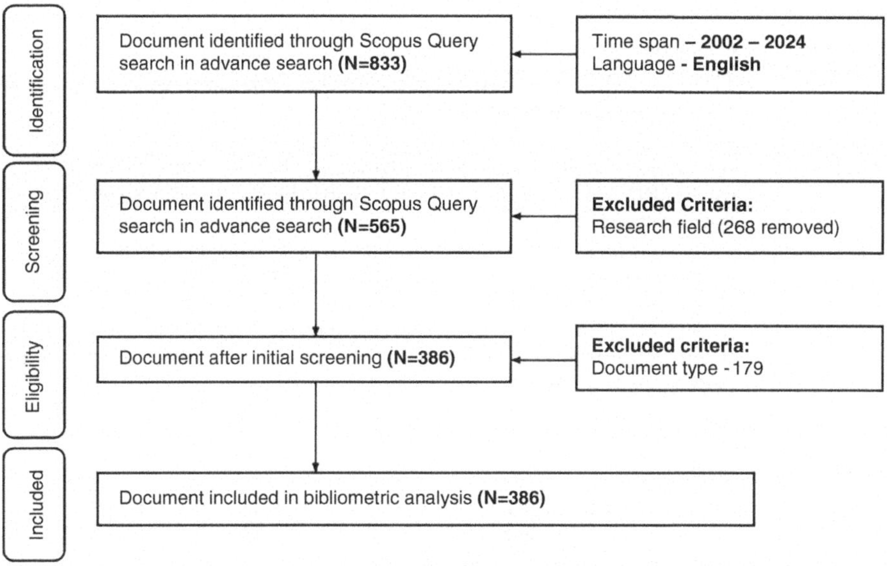

Fig. 1 PRISMA chart representing the collection of sustainable finance and environmental sustainability retrieved from Scopus database

information: Analytical tools like MS excel, Biblioshiny R, and VOSviewer were employed by the researchers to undertake the analysis.

4.2 Refined Search Query

To ensure capturing the most relevant papers, the authors performed an advanced search for documents where the title of the paper contains the primary keywords "green finance" and "environmental sustainability," along with the limitations and exclusion criteria provided in Table 1. The search was carried out with emphasis on the English language, peer-reviewed journal articles, and the time frame of the year 2002 up to and including the year 2024. Firstly, this search provided 833 papers. After using filters to select only publications in the Business, Management, Accounting, Social Sciences, Economics, Econometrics, and Finance fields, the number was brought down to 565 papers. Finally, articles with different publication dates and written in English were chosen for analysis; a total of 386 articles from journals of varying discipline areas.

The summary statistics depicted in Table 2 proclaim that a total of 386 articles (including articles and review articles) have been published in the Scopus database in the last 22 years (from the year 2002 to June 2024). All these documents contain an average of 19.45 citations each, and with an increase of 24.12% within 1 year. This is because the number of scholarly articles on environmental sustainability

Table 1 Search query description

Description	Conditions
Search query	TITLE-ABS-KEY (green AND finance AND environmental AND sustainability) **No of documents—833**
Search query after refining	TITLE-ABS KEY (green AND finance AND environmental AND sustainability) AND (LIMIT-TO (SUBJAREA, "ECON") OR LIMIT-TO (SUBJAREA, "SOCI") OR LIMIT-TO (SUBJAREA, "BUSI")) **No of documents—565**
Search query after final refining	TITLE-ABS-KEY ((green AND finance AND environmental AND sustainability) AND (LIMIT-TO (SUBJAREA, "ECON") OR LIMIT-TO (SUBJAREA, "SOCI") OR LIMIT-TO (SUBJAREA, "BUSI")) AND (LIMIT-TO (DOCTYPE, "ar")) AND (LIMIT-TO (PUBSTAGE, "final")) AND (LIMIT-TO (SRCTYPE, "j")) AND (LIMIT-TO (LANGUAGE, "English")) **No of documents—386**
Query search date	June 21st, 2024
Years	2002–2024
Subject area	Business, Management and Accounting, Social Sciences, Economics, Econometrics and Finance
Source type	The search was limited to Journals
Document type	Journal articles
Publication stage	Final publication
Language	The search was limited to the English language

Table 2 Summary statistics of the data used in this study

Sl. No.	Description	Results
1.	"Timespan"	2002:2024
2.	"Sources (Journals, Books, etc.)"	151
3.	"Documents"	386
4.	"Annual Growth Rate %"	24.12
5.	"Document Average Age"	1.91
6.	"Average citations per documents"	19.45
7.	"References"	22,896
8.	"Author's Keywords (DE)"	1141
9.	"Authors"	997
10.	"Authors of single-authored documents"	64
11.	"Single-authored documents"	66
12.	"Co-Authors per Documents"	3.05
13.	"International co-authorships %"	33.42

green finance has recently embraced a steepest slope as evidenced by the high average citation count per document. Furthermore, the findings revealed that 997 distinct authors have written papers in this field. Of this total, there are 66 single-authored documents, and on average, there are 3.05 co-authors per document, indicating that research in this field has more authors for each document as compared to others. Furthermore, the international co-authorship ratio of 33.42% also highlights this aspect.

4.3 *Performance Analysis*

In response to the first research question (RQ1), which centers on the annual growth rate of scientific production related to the concept of environmental sustainability and green finance, Fig. 2 presents the annual trend in scientific publications on environmental sustainability green finance as obtained from the Scopus database from 2002 up to 2024 with a compound annual growth rate of 24.12%. The COVID-19 has significantly contributed to more documents production from 2017 as 96% of articles produced are within the last 7 years, there is exponential growth in scientific production on green finance and environmental sustainability.

4.4 *Performance of Authors*

Table 3 further shows the 10 most prolific authors in terms of number and their affiliations. It is possible to uncover that Al-Faryan, M. A. S. from the Faculty of Business and Law at Portsmouth, United Kingdom, has written the five articles.

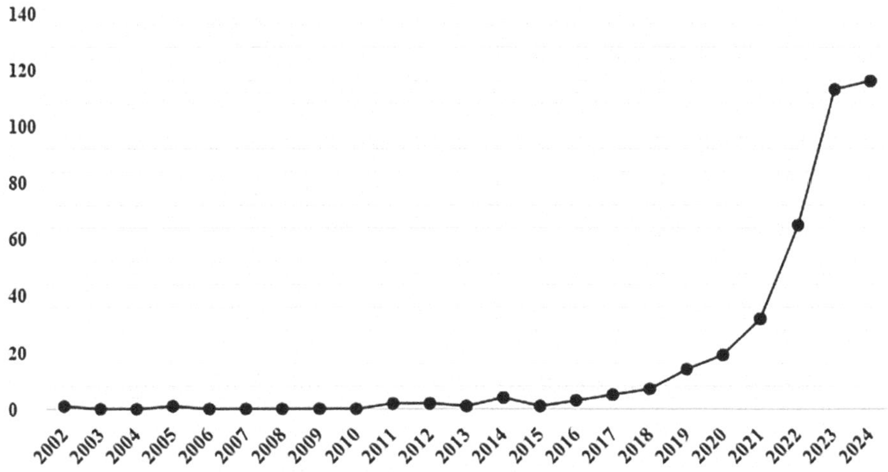

Fig. 2 Annual scientific production of sustainable finance and environment sustainability (1983–2024)

Table 3 Most productive authors

Rank	Authors	Affiliation	Articles
1.	Al-Faryan, MAS	Faculty of Business and Law, Portsmouth, United Kingdom	5
2.	Ibrahim, R.L.	University of Lagos , Lagos, Nigeria	4
3.	Lin, J.H.	Tamkang University, Taipei, Taiwan	4
4.	Sadiq, M.	Taylor's University Malaysia, Subang Jaya, Malaysia	4
5.	Udeagha, M.C.	University of Pretoria, Pretoria, South Africa	4
6.	Mohsin, M.	Taif University, Taif, Saudi Arabia	3
7.	Nassani, A.A.	King Saud University, Riyadh, Saudi Arabia	3
8.	Sharif, A.	Sunway University, Bandar Sunway, Malaysia	3
9.	Siddik, A.B.	University of Science and Technology of China, Hefei, China	3
10.	Taghizadeh-Hesary, F.	Tokai University, Tokyo, Japan	3

Close behind are Ibrahim, R. L from the University of Lagos, Lagos, Nigeria, with four papers followed by Lin, J. H from Tamkang University, Taipei, Taiwan, and Sadiq, M from Taylor's University Malaysia, Subang Jaya, Malaysia, also with four papers in the area of environmental sustainability green finance. It is evident that other authors within this research domain have produced one, two, or three articles at most.

4.5 *Performance of Documents*

Table 4 presents the top 10 sources that appear most frequently when it comes to environmental sustainability green finance domain. The most frequently cited paper is "Corporate Green Bonds" written by FLAMMER C and published in the Journal of Financial Economics in 2021, with 532 citations. Closely behind is a paper; "Environmental decentralization, digital finance and green technology innovation" written by Feng S, published in Structural Change and Economic Dynamics in 2022 having 305 citations. Another notable contribution is the paper titled "Green supply chain management and financial performance: One of the most cited papers which contributes to the same understanding is the mediating roles of operational and environmental performance" by Feng M with 209 citations retrieved from the database (Fig. 3).

4.6 *Performance of Journals*

From the total number of identified articles, it can be seen that the top 10 journals that have contributed immensely are responsible for 202 articles which is 52% of the total articles (Table 5). In fact, 46 journals (18.91%), which contain 73 articles,

Table 4 The 10 most cited articles

Rank	Paper	Title of the paper	DOI	Total citations	TC per year	Normalized TC
1	FLAMMER C, 2021, J FINANC ECON	"Corporate green bonds"	10.1016/j.jfineco.2021.01.010	532	133	10.86
2	FENG S, 2022, STRUCT CHANGE ECON DYN	"Environmental decentralization, digital finance and green technology innovation."	10.1016/j.strueco.2022.02.008	305	101.67	9.41
3	FENG M, et al. 2018, BUS STRATEGY ENVIRON	"Green supply chain management and financial performance: The mediating roles of operational and environmental performance."	10.1002/bse.2033	209	29.86	3.53
4	MADALENO M, et al. 2022, ENERGY ECON	"A step forward on sustainability: The nexus of environmental responsibility, green technology, clean energy and green finance."	10.1016/j.eneco.2022.105945	178	59.33	5.49
5	AGYABENG-MENSAH Y, et al. 2020, J CLEAN PROD	"Exploring financial performance and green logistics management practices: Examining the mediating influences of market, environmental and social performances."	10.1016/j.jclepro.2020.120613	152	30.4	3.56
6	NASSANI AA, et al. 2017, J CLEAN PROD	"Environmental Kuznets curve among BRICS countries: Spot lightening finance, transport, energy and growth factors."	10.1016/j.jclepro.2017.04.025	142	17.75	3.24
7	FALCONE PM, 2020, INT J GREEN ECON	"Environmental regulation and green investments: the role of green finance."	10.1504/ijge.2020.109735	141	28.2	3.3
8	KHAN MA, et al. 2022, BORSA ISTANB REV	"Does green finance really deliver what is expected? An empirical perspective."	10.1016/j.bir.2021.07.006	138	46	4.26
9	SADIQ M, et al. 2021, CHINA FINANCE REV INT	"Does green finance matter for sustainable entrepreneurship and environmental corporate social responsibility during COVID-19?"	10.1108/CFRI-02-2021-0038	125	41.67	3.86
10	REN X, et al. 2022, TECHNOL FORECAST SOC CHANGE	"The interrelationship between the carbon market and the green bonds market: Evidence from wavelet quantile-on-quantile method."	10.1016/j.techfore.2022.121611	121	40.33	3.73

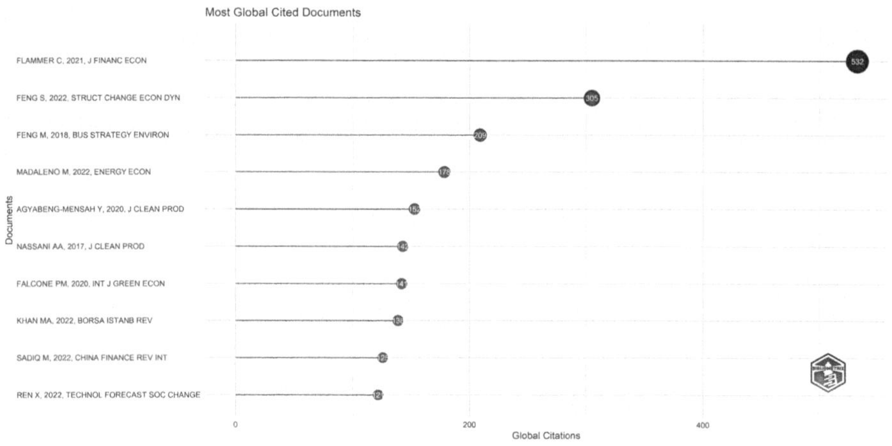

Fig. 3 The 10 most cited articles related to the sustainable finance and environmental sustainability

Table 5 Top 10 most relevant journals

Rank	Sources	IF	Articles	%	SNIP	SJR	Q	ABDC	Publisher
1	"Resources Policy"	13.4	77	20	2.083	2.063	Q1	B	Elsevier
2	"Sustainability (Switzerland)"	6.8	33	8.5	1.086	0.672	Q1	(NI)	MDPI
3	"Energy Economics"	18.6	25	6.4	2.637	3.555	Q1	A*	Elsevier
4	"Journal of Cleaner Production"	20.4	20	5.1	2.236	2.058	Q1	A	Elsevier
5	"Economic Research-Ekonomska Istrazivanja"	7.1	13	3.3	1.228	0.83	Q1	(NI)	Taylor & Francis
6	"Economic Change and Restructuring"	4.8	12	3.1	1.34	0.713	Q2	B	Springer Nature
7	"Environment, Development and Sustainability"	10.2	7	1.8	1.297	0.889	Q1	(NI)	Springer Nature
8	"Borsa Istanbul Review"	7.6	5	1.3	1.736	1.04	Q1	(NI)	Borsa Istanbul Anonim Sirketi
9	"Business Strategy and The Environment"	22.5	5	1.3	3.043	3.666	Q1	A	John Wiley & Sons
10	"Technological Forecasting and Social Change"	21.3	5	1.3	2.945	3.118	Q1	A	Elsevier

(NI)—Not indexed in ABDC list, IF—Impact factor, Q—Quartiles classification based upon Scopus database, and ABDC—Australian Business Dean's Council published only four, three, or only two articles, and the majority of journals, which are 111 articles or 29% of the journals, each published a single article related to environmental sustainability green finance. Currently, the highest representation on the topic is seen in "Resources Policy" with 77 green finance for environmental sustainability papers, which is about 20% of the total number of papers published in the journal; "Sustainability (Switzerland)" with 33 papers, Energy Economics with 25 papers, and Journal of Cleaner Production with 20 papers. Economic Research-Ekonomska Istrazivanja is in the second place with 13 ranks, The Economic Change and Restructuring, and Borsa Istanbul Review Business Strategy, The Environment Technological Forecasting & Social Change each has five ranks. As it is seen these top three tier journals together make the 35% of the articles on environmental sustainability green finance. Besides, the overall trend in this area of research has been characterized by a steep incline, especially after 2017, and a significant temporal gap has been observed (Fig. 4).

As it can be observed in Fig. 5, the patterns of publication sources in "environmental sustainability and green finance" vary from year to year although both "Resources Policy" and "Sustainability (Switzerland)" are published more frequently than other publications in each year. Furthermore, annual growth of such publications with the onset from 2021 suggests an exponential increase in disparity among sources.

Based on Bradford's law as presented in Fig. 6, it is clear that approximately 148 journals combined provide 251 articles in the inter collection on environmental sustainability and green finance with three core journals publishing 135 articles on this subject. The unification principle of Bradford's law is applied in the assessment of core journals within particular disciplines by accounting for the law of diminishing returns and scattering (Sadeqi-Arani & Kadkhodaie, 2023).

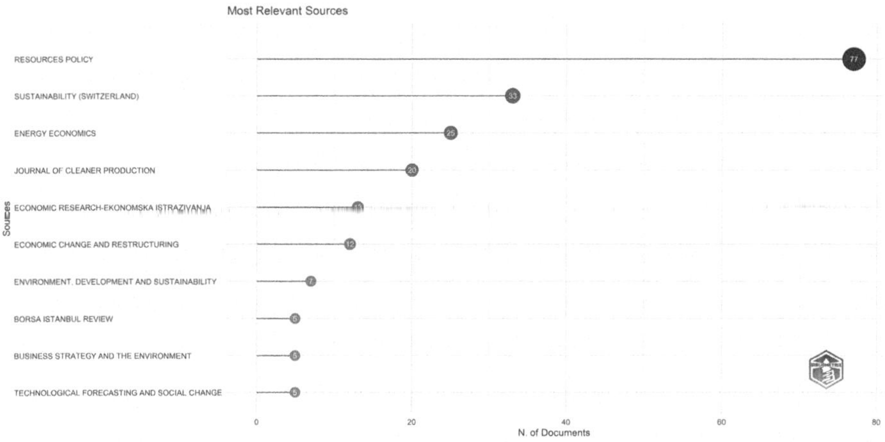

Fig. 4 Top 10 most relevant journals

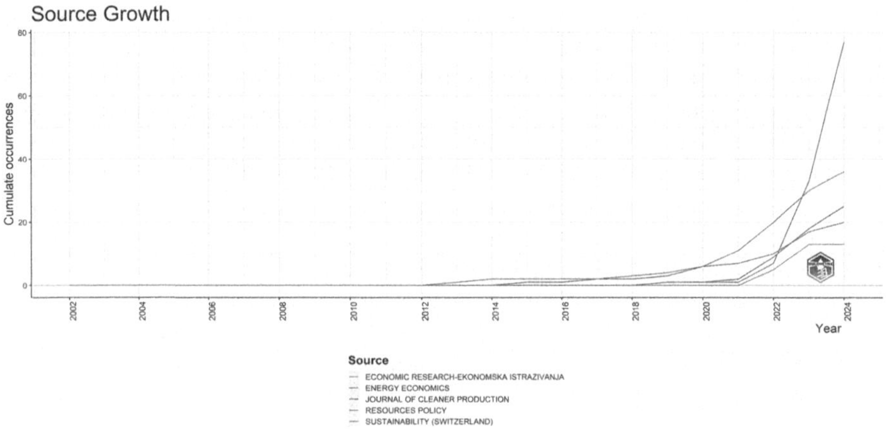

Fig. 5 The source dynamics of top 8 journals

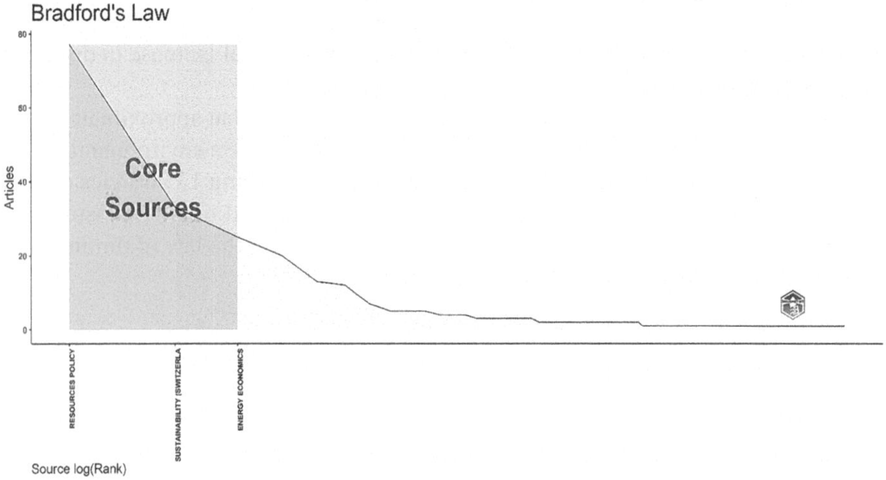

Fig. 6 Bradford's law of core sources of journals

Bradford law divides journals into three zones. The first third (Zone 1 or core) contains the cited journals most frequently used by researchers in the subject literature, which are commonly considered of the highest interest to the discipline's practitioners. The middle third of the journals (Zone 2) comprises the journal with moderate citation rate while the lower third of the journal citation or the third bottom (Zone 3 or tail) is the set of the journals that are rarely cited and still are on the margin of discursive formation of the particular discipline (K. Singh, 2014). According to Bradford's law, Zone 2 will contain n or an equal number of journals as those in Zone 1, while Zone 3 will contain n^2 times more journals than Zone 1 (Nash-Stewart et al., 2012).

4.7 *Performance of Countries*

Table 6 captures the articles published by different countries on environmental sustainability and green finance. China has led with 161 articles followed by India with 14 articles and Indonesia with 12 articles meaning that Asia has the largest share of articles published on the topic with 48%. 44% of all articles. These articles are categorized into two types based on author collaboration: as "Single Country Publications" (SCP) and "Multiple Country Publications" (MCP). In SCP, all authors of the articles are from the same country hence pointing to intracountry research collaboration. On the other hand, MCP can incorporate the authors from different countries, which may imply the inter-country or international collaborations (Sadeqi-Arani & Kadkhodaie, 2023). Among the top 19 countries, China can be noted as international collaborators, with 37% of MCP (Table 6) (Fig. 7).

Table 7 presents the extent of international cooperation of the research on environmental sustainability and green finance worldwide. The analysis of the data shows that China and Pakistan have published 21 articles together, China and Malaysia 13, and China and Saudi Arabia 11 articles, thus identifying strong cooperation in terms of research in green finance and environmental concerns (Refer to Fig. 8).

Table 6 —Top productive countries

Rank	Country	Articles	SCP	MCP	Freq	MCP_Ratio (%)
1	China	161	101	60	0.417	37
2	India	14	13	1	0.036	7
3	Indonesia	12	8	4	0.031	33
4	Italy	12	10	2	0.031	17
5	United Kingdom	12	10	2	0.031	17
6	Germany	11	9	2	0.028	18
7	Australia	9	6	3	0.023	33
8	USA	9	6	3	0.023	33
9	Saudi Arabia	8	2	6	0.021	75
10	South Africa	7	4	3	0.018	43
11	Spain	7	5	2	0.018	29
12	Pakistan	6	4	2	0.016	33
13	France	5	2	3	0.013	60
14	Malaysia	5	4	1	0.013	20
15	Turkey	5	1	4	0.013	80
16	Brazil	4	2	2	0.01	50
17	Portugal	4	3	1	0.01	25
18	Iran	3	2	1	0.008	33
19	Japan	3	1	2	0.008	67

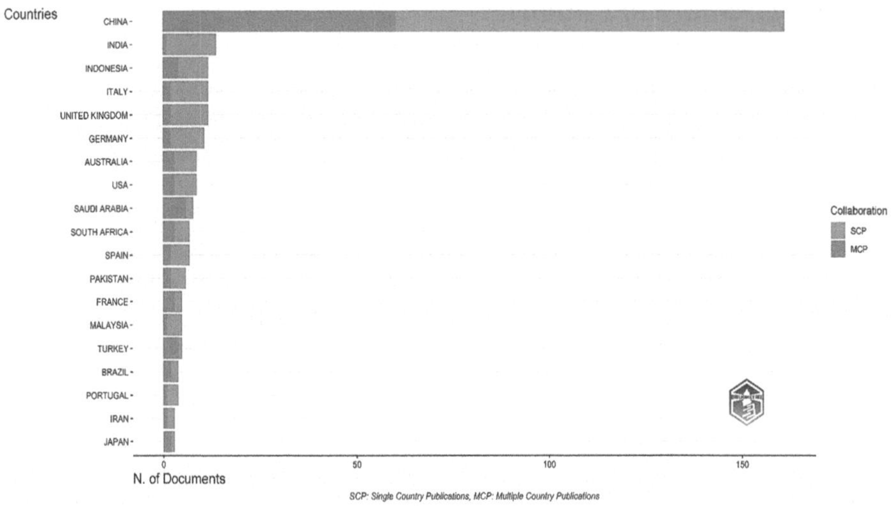

Fig. 7 Corresponding author's country

Table 7 Number of country collaboration around the world

Rank	From	To	Frequency
1	China	Pakistan	21
2	China	Malaysia	13
3	China	Saudi Arabia	11
4	China	United Kingdom	9
5	Pakistan	Malaysia	7
6	Pakistan	Saudi Arabia	7
7	China	Australia	6
8	China	Uzbekistan	5
9	China	Lebanon	4
10	China	Poland	4
11	China	USA	4
12	China	France	3
13	China	Ghana	3
14	China	Nigeria	3
15	China	Turkey	3
16	India	Turkey	3
17	Malaysia	Saudi Arabia	3
18	Malaysia	USA	3
19	Saudi Arabia	Jordan	3
20	Saudi Arabia	Uzbekistan	3

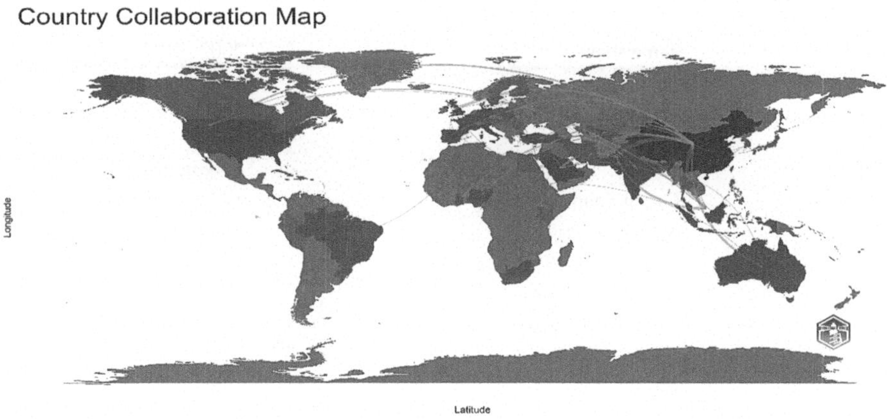

Fig. 8 Country scientific production and country collaboration map on the sustainable finance and environmental sustainability literature around the world

4.8 Conceptual Structure

To answer the third research question (RQ3) on possible gaps and key topics for further study in the existing literature on environmental sustainability and green finance environment, a word cloud analysis and co-word frequency analysis were performed. The summarized results are presented in the following figure and table, respectively:

Moreover, to answer the last research question (RQ4) related to the dynamics of the discussed field of research on environmental sustainability and green finance, word growth (year-wise) and temporal overlay on the co-word map of the literature were employed. The latter findings are also presented in summarized form in Table 8. The research in this study utilizes co-word analysis to present the conceptual layout of the terms selected and to highlight the key areas of research focus and possible gaps in the development of green finance and environmental sustainability domain.

It is graphically presented in Fig. 9, showing the top forty authors' keywords most often used in the study of environmental sustainability and sustainable finance. In our analysis, the following three terms were found to be among the most often used: "green finance" or "greening of finance" (13%), "sustainability" or "finance sustainability" (5%), and "environmental sustainability" (4%). It is possible to observe the dynamics of keywords in the field of green finance and environmental sustainability over 2002–2024 in Table 8. Other threads identified over this duration include: green finances, sustainable finances, sustainable development, green innovations, nature resources, green bonds, and sustainable development goals.

The KW cloud depicts the 40 most utilized keywords from the period between 2002 and 2024. In fact, a keyword cloud serves as graphic representation of the text data and the words with high weight coefficient are displayed larger. The results

Table 8 Frequency and evolution of keywords

	Time span:	2002–2024
	Documents: Sources: Authors:	386 151 997
Rank	**Terms**	**Frequency**
1	Green finance	127
2	Sustainability	46
3	Environmental sustainability	43
4	Sustainable finance	31
5	Natural resources	28
6	Fintech	27
7	China	25
8	Sustainable development	25
9	Climate change	23
10	Green bonds	18
11	Green innovation	16
12	Economic growth	14
13	ESG	14
14	Financial development	14
15	Renewable energy	14
16	Financial performance	12
17	Co₂ emissions	11
18	Environmental performance	11
19	Corporate governance	10
20	Sustainable development goals	10
21	Corporate social responsibility	9
22	Ecological footprint	9
23	Green technology innovation	9
24	Green bond	8
25	Green technology	8
26	BRICS	7
27	Finance	7
28	Green economy	7
29	Green energy	7
30	Green financing	7
31	Bibliometric analysis	6
32	Carbon emission	6
33	Climate finance	6
34	Covid-19	6
35	Energy transition	6
36	Environmental protection	6
37	Financial inclusion	6

(continued)

Table 8 (continued)

	Time span:	2002–2024
38	Greenwashing	6
39	Impact investing	6
40	mmqr	6

Fig. 9 Word cloud

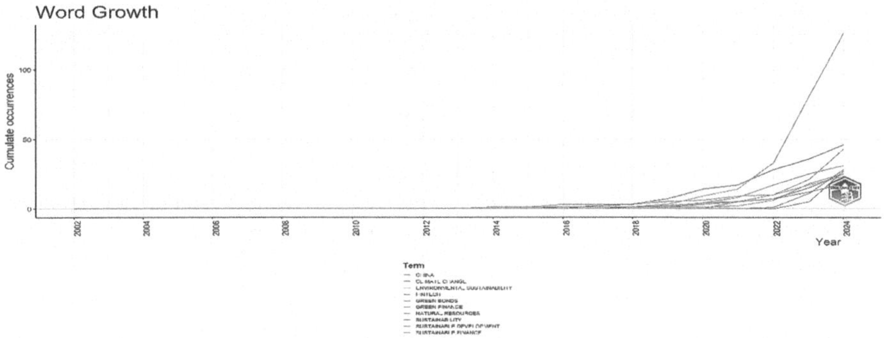

Fig. 10 Word growth (year-wise) of top 10 frequent keywords used in publications

reveal that the most frequently indicated keywords are "green finance," "environmental sustainability," "sustainability," "sustainable finance," and "climate change." Further, "green financing," "green technology innovation," "green technology," "energy transition," "green economy," and "climate finance" are listed as areas that have been undertaken with less focus. The annual frequency of the top 10 most commonly utilized keywords is depicted in Fig. 10, demonstrating that the occurrence of certain terms has steadily grown. In particular, "green finance" has developed rapidly, especially a 23-fold increase in average from year 2002 to year 2024.

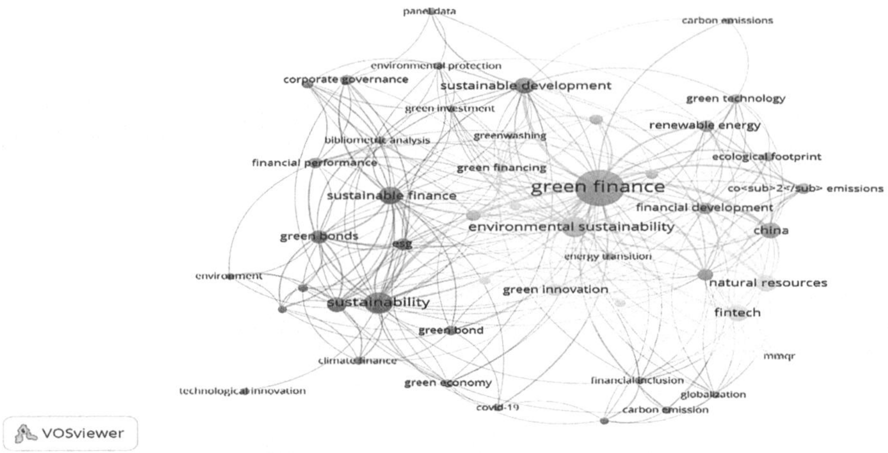

Fig. 11 Network of co-word occurrence (threshold 5 occurrences, display 48 keywords)

4.9 Co-words Analysis

Co-word analysis involving "the simultaneous occurrence of two words to elucidate the concept and content of a scientific discipline" (Callon et al., 1983). Co-occurrence is described by a statistic measure like correlation or mutual information in this method. Indeed, co-word analysis helps uncover relationships between topics within a research domain by comparing the frequency of simultaneous usage of pairs of words or phrases and capture scientific progress (He, 1999). It also becomes a useful method for defining and illustrating the relations between the keywords within the selected scientific field (Del Barrio-García et al., 2020).

Figure 11 shows the network map that depicts the co-word frequency of environmental sustainability and sustainable finance in scholarly literature. Thus, the threshold has been set at 5 and within the frame of 48 keywords, the research reveals six significant clusters within this domain. To do this, the size of each circle is proportional to the usage frequency of each keyword; meanwhile, each cluster is depicted in a unique color. All the words pertaining to a cluster have the same color whereby, the related topics in each cluster are visually emphasized.

Figure 12 presents the thematic cluster map derived from the literature about environmental sustainability and sustainable finance with the minimum count of 5 and overlaying the time line extracted from authors' keywords resulting 48 themes. The size of each circle represents their frequency of usage of the said keyword. The circle's color depicts the average year of co-occurrence, which has been mentioned in the legend at the lower right of the figure. Specifically, the temporal overlay shows that "financial performance," "corporate governance," "corporate social responsibility," "environmental protection," and "climate finance" (highlighted in purple) were investigated earlier. On the other hand, topics such as "fintech,"

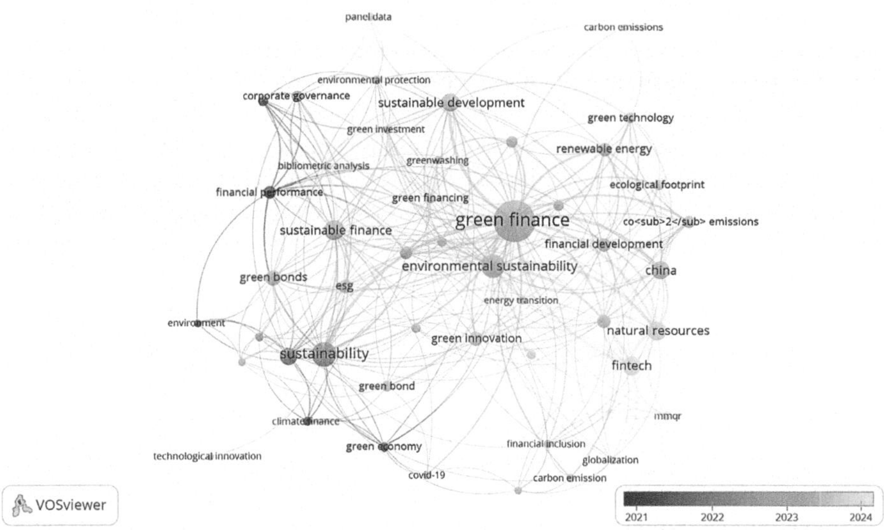

Fig. 12 Temporal overlay on a co-word map (threshold 5 occurrences, display 48 keywords)

"mmqr," "natural resources," "BRICS," and "bibliometric analysis" have emerged rather recently.

5 Conclusion

Thus, this article provides descriptive data on environmental sustainability and sustainable finance, specifically, covering the aspects of the most and least represented countries, the dynamics of the annual publications, authors' productivity, articles' citation rates, and source journals. The bibliometric analysis enhances the literature by revealing the research's conceptual framework and potential growth trajectory using the co-word approach.

A total of 386 scientific publications from the Scopus database spanning 2002–2024 were analyzed, revealing exponential growth in the past 6 years, with over 93% of articles published between 2019 and 2024. Leading contributors include China, India, and Indonesia. Notably, "Resources Policy" leads with 77 papers, followed by "Sustainability (Switzerland)" with 33 articles, and "Energy Economics" with 25 papers, among others. The top 10 authors, constituting 1.0% of total authors, contributed 36 articles, representing 9.32% of total publications.

The most cited articles include "Corporate Green Bonds" by Flammer C. (2021) with 532 citations, followed by "Environmental decentralization, digital finance and Green Technology Innovation" by Feng S. (2022) with 305 citations, and "Green Supply Chain Management and financial performance: The most cited article is The mediating roles of operational and environmental performance" by Feng et al. (2018) with a total of 209 citations proving their importance for the topic at hand.

China is ranked at the top of SCP countries and MCP countries showing active engagement of these countries through international collaboration in green finance and environmental sustainability research.

This research advances the literature by providing the historical trend analysis of themes such as "green finance," "fintech," "sustainable development," "green innovation," "natural resources," "climate change," "green bonds," and "sustainable finance" for the period 2021–2024. A keyword co-occurrence analysis reveals fundamental areas of significance in the field. It proves its efficiency in identifying new trends and directions in a given subject area and supporting the knowledge discovery database process.

6 Limitations and Future Direction of the Study

The present study's findings contribute to the existing body of knowledge on environmental sustainability and sustainable finance in enabling researchers of different sectors. However, due to the nature of this study as a bibliometric analysis, it also has some limitations. It only uses data from the Scopus database, indicating that future studies could improve the comprehensiveness by using other database sources too like Google Scholar, WOS, and Dimension. Further, this research centers on 386 selected journal articles in Business, Management, Accounting, Social Sciences, Economics, Econometric, and Finance fields. This research could be expanded in future to also cover other disciplines in order get even more accurate picture of the publication trends in these areas. Additionally, future studies should extend the work by identifying green investment opportunities that have reduced carbon intensity and increase the participation of underrepresented groups. Further, that more opportunities can be pursued in fields such as social and climate finance projects that can be inaugurated in the underdeveloped nations emitting high carbon emission rates. This study employed R programming version 4.2.2 and VOSviewer version 1.6.20 to generate visualization maps for data analysis and visualization. Future researchers may consider employing alternative software tools such as Jamovi, N-vivo, Gephi, HistCite, BibExcel, or Publish or Perish for their analyses. These tools can provide additional perspectives on the data, enhancing the understanding of environmental sustainability and contributing to the literature on sustainable finance.

References

Agyabeng-Mensah, Y., Afum, E., & Ahenkorah, E. (2020). Exploring financial performance and green logistics management practices: Examining the mediating influences of market, environmental and social performances. *Journal of Cleaner Production, 258*, 120613. https://doi.org/10.1016/j.jclepro.2020.120613

Ahmi, A., & Nasir, M. H. M. (2019). Examining the trend of the research on eXtensible Business Reporting Language (XBRL): A bibliometric review. Social Science Research Network. https://papers.ssrn.com/sol3/Delivery.cfm/SSRN_ID3839843_code988359.pdf?abstractid=3839843&mirid=1

Callon, M., Courtial, J., Turner, W. A., & Bauin, S. (1983). From translations to problematic networks: An introduction to Co-word analysis. *Social Science Information, 22*(2), 191–235. https://doi.org/10.1177/053901883022002003

Chang, B. H., Auxilia, P. A. M., Kalra, A., Wong, W., & Uddin, M. A. (2023). Greenhouse gas emissions and the rising effects of renewable energy consumption and climate risk development finance: Evidence from BRICS countries. *Annals of Financial Economics, 19*(02), 2350007. https://doi.org/10.1142/s2010495223500070

Chen, X., & Chen, Z. (2021). Can green finance development reduce carbon emissions? Empirical evidence from 30 Chinese provinces. *Sustainability, 13*(21), 12137. https://doi.org/10.3390/su132112137

Del Barrio-García, S., Muñoz-Leiva, F., & Golden, L. (2020). A review of comparative advertising research 1975–2018: Thematic and citation analyses. *Journal of Business Research, 121*, 73–84. https://doi.org/10.1016/j.jbusres.2020.08.023

Ellegaard, O., & Wallin, J. A. (2015). The bibliometric analysis of scholarly production: How great is the impact? *Scientometrics, 105*(3), 1809–1831. https://doi.org/10.1007/s11192-015-1645-z

Falcone, P. M. (2020). Environmental regulation and green investments: The role of green finance. *International Journal of Green Economics, 14*(2), 159. https://doi.org/10.1504/ijge.2020.109735

Falcone, P. M., & Sica, E. (2019). Assessing the Opportunities and Challenges of green finance in Italy: An analysis of the biomass production sector. *Sustainability, 11*(2), 517. https://doi.org/10.3390/su11020517

Fang, Y., & Shao, Z. (2022). How does green finance affect cleaner industrial production and end-of-pipe treatment performance? Evidence from China. *Environmental Science and Pollution Research International, 30*(12), 33485–33503. https://doi.org/10.1007/s11356-022-24513-4

Feng, M., Yu, W., Wang, X., Wong, C. Y., Xu, M., & Xiao, Z. (2018). Green supply chain management and financial performance: The mediating roles of operational and environmental performance. *Business Strategy and the Environment, 27*(7), 811–824. https://doi.org/10.1002/bse.2033

Feng, S., Zhang, R., & Li, G. (2022). Environmental decentralization, digital finance and green technology innovation. *Structural Change and Economic Dynamics, 61*, 70–83. https://doi.org/10.1016/j.strueco.2022.02.008

Flammer, C. (2021). Corporate green bonds. *Journal of Financial Economics, 142*(2), 499–516. https://doi.org/10.1016/j.jfineco.2021.01.010

Gu, X., Qin, L., & Zhang, M. (2023). The impact of green finance on the transformation of energy consumption structure: Evidence based on China. *Frontiers in Earth Science, 10*, 1097346. https://doi.org/10.3389/feart.2022.1097346

He, Q. (1999). Knowledge Discovery through Co-Word Analysis. *Library Trends, 48*(1), 133–159. https://dblp.uni-trier.de/db/journals/libt/libt48.html#He99

Huynh, T. L. D., Walther, T., & Utz, S. (2022). Green and Sustainable finance in the Asia-Pacific Markets: An introduction to the special issue. *Asia-Pacific Financial Markets, 29*(1), 1–3. https://doi.org/10.1007/s10690-022-09362-y

Khan, H. H., Ahmad, N., Yusof, N. M., & Chowdhury, M. A. (2024). Green finance and environmental sustainability: A systematic review and future research avenues. *Environmental Science and Pollution Research, 31*(6), 9784–9794. https://doi.org/10.1007/s11356-023-31809-6

Khan, M. A., Riaz, H., Ahmed, M., & Saeed, A. (2022). Does green finance really deliver what is expected? An empirical perspective. *Borsa Istanbul Review, 22*(3), 586–593. https://doi.org/10.1016/j.bir.2021.07.006

Koczar, J., Vagizova, V., Agliullina, M., & Agliullina, Z. (2020). Application of green finance tools in the system of sustainable interaction between the real and financial economy: Opportunities and prospects. Proceedings of the "New Silk Road: Business Cooperation and Prospective of Economic Development" (NSRBCPED 2019). https://doi.org/10.2991/aebmr.k.200324.113.

Li, C., Fan, X., Hu, Y., Yan, Y., Shang, G., & Chen, Y. (2022). Spatial spillover effect of green finance on economic development, environmental pollution, and clean energy production across China. *Environmental Science and Pollution Research International, 29*(58), 87858–87873. https://doi.org/10.1007/s11356-022-21782-x

Madaleno, M., Dogan, E., & Taskin, D. (2022). A step forward on sustainability: The nexus of environmental responsibility, green technology, clean energy and green finance. *Energy Economics, 109*, 105945. https://doi.org/10.1016/j.eneco.2022.105945

Meo, M. S., & Karim, M. Z. A. (2022). The role of green finance in reducing CO2 emissions: An empirical analysis. *Borsa Istanbul Review, 22*(1), 169–178. https://doi.org/10.1016/j.bir.2021.03.002

Muganyi, T., Yan, L., & Sun, H. (2021). Green finance, fintech and environmental protection: Evidence from China. *Environmental Science & Ecotechnology, 7*, 100107. https://doi.org/10.1016/j.ese.2021.100107

Nash-Stewart, C. E., Kruesi, L. M., & Del Mar, C. B. (2012). Does Bradford's law of scattering predict the size of the literature in Cochrane reviews? *Journal of the Medical Library Association: JMLA, 100*(2), 135–138. https://doi.org/10.3163/1536-5050.100.2.013

Nassani, A. A., Aldakhil, A. M., Abro, M. M. Q., & Zaman, K. (2017). Environmental Kuznets curve among BRICS countries: Spot lightening finance, transport, energy and growth factors. *Journal of Cleaner Production, 154*, 474–487. https://doi.org/10.1016/j.jclepro.2017.04.025

Nawaz, M. A., Seshadri, U., Kumar, P., Aqdas, R., Patwary, A. K., & Riaz, M. (2020). Nexus between green finance and climate change mitigation in N-11 and BRICS countries: empirical estimation through difference in differences (DID) approach. *Environmental Science and Pollution Research International, 28*(6), 6504–6519. https://doi.org/10.1007/s11356-020-10920-y

Peng, J., & Zheng, Y. (2021). Does environmental policy promote energy efficiency? Evidence from China in the context of developing Green Finance. *Frontiers in Environmental Science, 9*, 733349. https://doi.org/10.3389/fenvs.2021.733349

Rasoulinezhad, E., & Taghizadeh-Hesary, F. (2022). Role of green finance in improving energy efficiency and renewable energy development. *Energy Efficiency, 15*(2), 14. https://doi.org/10.1007/s12053-022-10021-4

Ren, X., Li, Y., Yan, C., Wen, F., & Lu, Z. (2022). The interrelationship between the carbon market and the green bonds market: Evidence from wavelet quantile-on-quantile method. *Technological Forecasting & Social Change/Technological Forecasting and Social Change, 179*, 121611. https://doi.org/10.1016/j.techfore.2022.121611

Ren, X., Shao, Q., & Zhong, R. (2020). Nexus between green finance, non-fossil energy use, and carbon intensity: Empirical evidence from China based on a vector error correction model. *Journal of Cleaner Production, 277*, 122844. https://doi.org/10.1016/j.jclepro.2020.122844

Roig-Tierno, N., Gonzalez-Cruz, T. F., & Llopis-Martinez, J. (2017). An overview of qualitative comparative analysis: A bibliometric analysis. *Journal of Innovation & Knowledge/Journal of Innovation and Knowledge, 2*(1), 15–23. https://doi.org/10.1016/j.jik.2016.12.002

Sadeqi-Arani, Z., & Kadkhodaie, A. (2023). A bibliometric analysis of the application of machine learning methods in the petroleum industry. *Results in Engineering, 20*, 101518. https://doi.org/10.1016/j.rineng.2023.101518

Sadiq, M., Nonthapot, S., Mohamad, S., Keong, O. C., Ehsanullah, S., & Iqbal, N. (2021). Does green finance matter for sustainable entrepreneurship and environmental corporate social responsibility during COVID-19? *China Finance Review International, 12*(2), 317–333. https://doi.org/10.1108/cfri-02-2021-0038

Schotten, M., El Aisati, M., Meester, W. J., Steiginga, S., & Ross, C. A. (2017). A brief history of Scopus: The world's largest abstract and citation database of scientific literature. In *Research analytics* (pp. 31–58). Auerbach Publications. https://doi.org/10.1201/9781315155890-3

Sharif, M. H. M., Ahmi, A., & Rosli, K. (2020). Measuring the trend of the research on social media in SMEs: A Bibliometric evaluation. *TEST Engineering & Management, 82*, 6437–6446. http://testmagzine.biz/index.php/testmagzine/article/view/1861

Singh, K. (2014). Application of Bradford's law on journal citations: A study of Ph. D. Theses [Doctoral dissertation].

Sinha, A., Mishra, S., Sharif, A., & Yarovaya, L. (2021). Does green financing help to improve environmental & social responsibility? Designing SDG framework through advanced quantile modelling. *Journal of Environmental Management, 292*, 112751. https://doi.org/10.1016/j.jenvman.2021.112751

Wu, X., Sadiq, M., Chien, F., Ngo, Q., Nguyen, A., & Trinh, T. (2021). Testing role of green financing on climate change mitigation: Evidences from G7 and E7 countries. *Environmental Science and Pollution Research International, 28*(47), 66736–66750. https://doi.org/10.1007/s11356-021-15023-w

Zakari, A., & Khan, I. (2022). The introduction of green finance: a curse or a benefit to environmental sustainability? *Energy Research Letters, 3*(3). https://doi.org/10.46557/001c.29977

Zhang, K. Q., Chen, H. H., Tang, L. Z., & Qiao, S. (2022). Green Finance, Innovation and the Energy-Environment-Climate Nexus. *Frontiers in Environmental Science, 10*, 879681. https://doi.org/10.3389/fenvs.2022.879681

Zyoud, S. H., Sweileh, W. M., Awang, R., & Al-Jabi, S. W. (2018). Global trends in research related to social media in psychology: mapping and bibliometric analysis. *International Journal of Mental Health Systems, 12*(1), 1–8. https://doi.org/10.1186/s13033-018-0182-6

The Impact of Government Initiatives on Sustainable Business Practices

Joice Ann Mathew and Susnato Maji

Abstract This paper examines the evolving role of government initiatives in fostering sustainable business practices, with a focus on Corporate Social Responsibility (CSR) as a critical tool for aligning economic growth, social equity, and environmental sustainability. By analyzing the historical evolution of CSR and its current framework in India, including its mandatory implementation under the Companies Act, 2013, this study highlights the multidimensional aspects of CSR: environmental, ethical, philanthropic, and economic responsibilities. It also explores the strategic integration of CSR to meet stakeholder expectations, enhance competitiveness, and mitigate risks, illustrating its potential as a driver for sustainable development. The research further addresses mechanisms for legal compliance, inspection, and enforcement, offering insights into the challenges and opportunities for businesses in adhering to CSR norms while contributing to broader societal goals.

1 Introduction

Being able to sustain a business practice in the long run through economic efficiency, social appropriateness, and environmentally sensitivity is regarded as sustainable business practice (IBM, 2021). Sustainable business practices involve business exercises performed in a manner that does not harm the environment or involve the use of resources in a way that contributes to their depletion. It is further described as an attempt to integrate economic prosperity or financial development with sustainable resources and social equity. Some of the action areas are emission reductions, utility of renewable energy, labor rights, and sustainable sourcing among others.

J. A. Mathew · S. Maji (✉)
Christ University, Pune, India

A. Sharma et al. (eds.), *The Synergistic Effect of Sustainable Business Practices on Corporate Performance*, https://doi.org/10.1007/978-981-96-3375-3_8

2 Corporate Social Responsibility

Corporate Social Responsibility (CSR) is a strategic framework for management to enable a company to evaluate its operations, decisions, actions, and policies, in order to ensure that it takes responsibility for social issues, which affect not only business organizations, but the public as well (Windsor, 2001). The implementation of CSR activities also allows to deliberate on all aspects of the society where the given company is involved: economic, social, and environmental. CSR is the set of activities that an organization performs in order to reflect the public good, and the law itself will never ask for this as a minimum standard.

2.1 Evolution of CSR

The concept of Corporate Social Responsibility cannot be said to have started recently as its existence can be traced back to earlier times. It is important to understand that CSR is not a new concept, but rather has a rich history that has transformed over time as people and markets have required more from corporations. CSR's first origins go back to early twentieth century when with industrialization and corporate development, called "social issues" began to emerge enormously affecting the society and the environment. The first observations about this concept are that it was initially linked to philanthropy and charity correct, with key businesspersons like the Carnegie of Rockefeller intensely practicing philanthropy and charity accept as true (Harvey et al., 2011).

However, another important aspect of CSR and its development toward recognition of companies' responsibility for the outcomes of their activities began an evolution in the mid-twentieth century. The history of CSR can be traced back to the work of Howard Bowen, an economist who published "Social Responsibilities of the Businessman" in the 1950s considered the cornerstone of the modern theories of CSR. Bowen emphasized the perspective, which can be summarized as the key social responsibilities that were stated as doing those things that are beneficial from the business point of view (Bowen, 1953). The subsequent decade, i.e. the 1960s and the 1970s were also marked by development of what is considered as CSR. Meanwhile, there emerged several social movements that included those that championed for civil rights, environmentalism, and consumerism. This created a field of unfettered exploration and enquiry of corporate behavior in practice.

The desire for the ethical conduct of business what was required in the 1980s and 1990s was borne out of globalization and corporate fraud such as the WorldCom and Enron frauds (O'brien, 2005). Business understood that the corporate value, sustainability, the specifics of the companies' activity, reputation, and endurance depended on the social and ecological criteria. This era also witnessed the creation of standards and frameworks that enhance CSR, including the United Nations

Global Compact as well as the Triple Bottom Line approach, which encompasses people, planet, and profit (Conway, 2018).

As new business entered the twenty-first century, CSR had already became a focal business practice due to awareness among business and their stake holders. Today, CSR summarizes components of environmental interaction, ethical employment, social contribution, and organizational management. This evolution continues as business organizations strive to incorporate better sustainability at the heart of their business strategies and activities and thus discover that exemplified eco-responsibility is actually the bottom line on sustainable business success and human welfare on the planet (Table 1).

2.2 *Parts of CSR*

Each CSR dimension is significantly essential for the sustainability of the business and its endeavor to deliver the benefits and improvements for the social environment in a responsible manner.

Table 1 Year-wise events and description that lead evolution of CSR

Year	Event	Description
1953	Howard Bowens' 'Social Responsibilities of the Businessman'	This book is often credited with laying the foundation for modern CSR by arguing that businesses have obligations to society beyond profit-making.
1970s	CSR gains traction with environmental and social movements	The rise of environmental and social movements during this decade, such as the establishment of Earth Day and growing awareness of social justice issues, brought attention to the social responsibilities of businesses.
1980s	Emergence of corporate codes of conduct	Companies began developing and adopting formal codes of conduct to guide their operations in ethical and socially responsible ways, this period also saw the rise of stakeholder theory.
1990s	Globalization and corporate scandals drive CSR focus	Increased globalization and high-profile corporate scandals highlighted the need for greater corporate accountability and transparency, leading to a stronger emphasis on CSR.
2000	Introduction of Global Reporting Initiative (GRI)	The GRI was launched, providing a comprehensive framework for companies to report on their environmental, social, and governance (ESG) performance.
2010	Increased emphasis on sustainability and ESG criteria	Businesses began integrating sustainability into their core strategies, and investors started considering ESG criteria as part of their investment decisions.
2020	Integration of CSR into Core business strategies	CSR evolved from being a peripheral activity to a central component of business strategy, with companies increasingly recognizing that sustainable practices are essential for long-term success.

2.2.1 Environmental Responsibility

Environmental responsibility categorizes under CSR since it refers to the corporation's interaction with the environment. It addresses the issues of environmental degradation caused by business ventures. Some of the activities may include measures being taken to reduce the emission of CO2, efficient management of wastes together with the conservation of natural resources. The activities may include conversion to green energy, cutting down on energy and water usage, lowering its emission rates, and having strong recycling policies. Farther, practically in tackling environmental issues, businesses do not only follow legal provisions but also participate in the discussion of the world about preserving species and fighting against global warming to leave a better planet for generations to come (Nie et al., 2019).

2.2.2 Ethical Responsibility

Ethical obligation is a form of operation of different companies to favorableness, integrity and something that respects the rights of every shareholder. The activities include compliance with the law requirement in terms of ethics, meeting the acceptable standard of ethical requirement, and enabling honesty. Hence, ethical practices are also treated as labor practices and non-discriminating allowing fair trade, and observing human rights along the supply furnish (Maloni & Brown, 2006). Ethical responsibilities which are implemented help organizations to build confidence within the environment or its stakeholders, enhance on its image, and avoid legal repercussions or tarnished reputation.

2.2.3 Philanthropic Responsibility

Philanthropic responsibility is the proactive activity that the firms undertake, with a view of making the surrounding environment a better place even in the absence of any external pressure. This includes gifting, sponsorships, and fund raising activities which are mostly related to social responsibilities like education, health, and poverty reduction. Some of such engagements represent the company as being involved in some kind of social responsibility beyond being responsible for profit making through it primary or main business.

2.2.4 Economic Responsibility

Economic responsibility embraces activities necessary for economic growth in the areas where operations are being conducted. This means the generation of employment, the fostering of new enterprises within the region, and the established turnover in business operations which create worth that is duly apportioned among various players. In this it can mean offering responsible investments by a company or support of small- and medium-sized enterprises or ensuring employees have fair wages and conditions. In this regard, businesses support economic sustainability in societies by creating favorable climates for the development and success in enterprise (Kitzmueller & Shimshack, 2012, p. 8).

2.3 Criteria for Firms to Implement CSR

CSR in India is not mandatory for all firms but it is mandatory for the selected companies those that fulfill the stipulated criteria of financial requirements of the Companies Act of 2013. While legal requirement pertains only to qualifying firms, most other firms are already engaging in CSR provision as an act of voluntarism for strategic management, benefits of stakeholders, increase in competitiveness, and effective risk management.

2.3.1 Legal Requirements

In India, CSR has been made compulsory for certain classes of companies under and as per law laid down by Companies Act, 2013. The provisions of this act especially Section 135 provides that depending on the size and profitability of companies that it applies to, then they should spend not less than 2% of their average net profits earned from the immediate three financial years in CSR. These activities are pointed out in Schedule VII of the Act which comprises areas of education, poverty annihilation, health, climate change- sustainability, and rural development. In addition, companies are required to establish CSR committees to supervise such programs with reference to the legal requirements.

2.3.2 Strategic Alignment

Since many supermarket moguls admit that CSR in India is a strategic and integral business priority, it means that they employ it for its own sake and not simply to meet expectations. CSR is incorporated into the general business objectives and business models to broaden the corporate reputation and consumer trust besides promoting sustainability in the longer perspective (Iglesias et al., 2020). As research shows, through CSR policies organizations can solve social and ecological issues that are of interest to the organization and its stakeholders. This strategic integration of CSR into the strategic plan guarantees that, CSR activities are not philanthropic activities but part of the strategies that the company uses to deliver shareholder value while also creating positive societal impacts.

2.3.3 Stakeholder Expectations

The second set of factors, namely stakeholder pressure, directly affects the adoption of CSR by firms across India. The key stakeholders such as customers, investors, employees, and the society have shifted their expectations of organizations to the side of the envelope and want organizations to be socially responsible and contribute to society. Satisfying these expectations benefits the companies by promoting positive perceptions from its stakeholders, higher employee satisfaction and hence better turnover rates, and greater attractiveness to investors who are socially responsible (Branco & Rodrigues, 2006). From the stakeholder's perspective, firms can be able to respond to stakeholder's expectations hence improve and gain the social license to operate hence the importance of creating a good public image in a competitive market.

2.3.4 Competitive Advantage

In the Indian context, CSR can indeed be strategically used as a competitive weapon where it can benefit firms and distinguish them from other players in the market. Firms that embrace CSR activities can set themselves apart from rivals in anticipation of occasions where they will be called upon to explain their stance on social and/or environmental matters (Smith, 2007). This can translate into higher customer retention and creation of a brand wall through the optimization of talent attraction and retention. In addition, CSR helps companies to obtain new and significant opportunities for business; for example, opportunities to gain access to markets which are concerned with sustainability and ethical standards, and opportunities for continued innovation by driving companies to come up with new products and services which are more sustainable.

2.3.5 Risk Management

Implementing CSR activities contributes to various risks management that a firm in India faces in terms of legal risks, companies' reputational risks and operational risks. By reverting to the CSR regulations and carrying out business activities that are free from unadulterated vices, companies can save themselves from the legal consequences that often accrue to unscrupulous business people and the business lose that they are likely to incur due to negative media publicity (McBarnet, 2009). In addition, strategic embracing of CSR commitments can enable organizations to come up with an early estimate of social and environmental risks that may affect the company. Thus, the goal of managing risk through CSR can increase organizational resilience, investment in business continuity, and community trust (Blundo et al., 2021).

3 Tackling Legal Liabilities and CSR Compliance Inspection and Enforcement in India

Consistent with other nations, The Companies Act, 2013 is the legal framework that governs CSR compliance in India with particular reference to Section 135. Companies should adhere to these requirements by meeting the following policies that have been developed and set by the Ministry of Corporate Affairs (MCA). The framework and enforcement mechanisms have been formulated and established. Those companies that exceed the limit of net worth or turnover or net profit must of their own accord expend not less than 2% of the total average net profit of the said related company that was made in respect of the 3 years preceding the aforementioned fiscal year.

In order for a state to have a sense of whether companies are using their money wisely, and for the general public to see transparency on how their money is spent on CSR, a new system has been developed: self-disclosure, compliance reporting, and audit. It will require the formation of a CSR Committee that shall develop the CSR Policy, and the Board will be required to monitor and facilitate its implementation. The activities performed shall also be reported annually in the Board's Report, as well as be placed on the company's website. These reports are then passed to the Ministry of Corporate Affairs, here the reports shall be scrutinized and looked at to see if they are compliant. It might also grant the RoC (Registrar of Companies) with inspection and scrutiny powers of the book of accounts to ensure the accuracy of the report is true. If it observes some features that are different from a similar case or a number of violations of legislation, the RoC may transfer the case to the inspection and investigation wing.

According to the provision of the law, it is mandatory that in each financial year, if the firm have not spent the requisite amount on CSR activities then they will have to record their explanation why they have not done so in the annual report. It is also important to view some of the penalties given for non-compliance. Pursuant to the Section 405 of the Companies (Amendment) Act, 2019, in case the company fails to expend the mandatory amount, then it will attract a minimum penalty of INR 50,000 to a maximum of INR 25,00,000. Any officer of the company who is in default shall be punishable with a fine which shall not be less than fifty thousand rupees or more than the amount equivalent to 500,000 rupees, or with imprisonment for a term not exceeding 3 years or with both. It also highlights the fact that CSR spending would be utilized as an instrument of accountability to ensure that spending becomes an instrument of change for the companies to embrace the concept as a tool of transforming their mentality and doing good to society rather than viewing it as an extra regulatory tool.

This combination of regulatory scrutiny and penal action under law ensures that the firms take their CSR commitments seriously. This framework would help create a culture of corporate responsibility and sustainable development in the Indian corporate sector, making business more meaningful to society.

4 Companies in India Actively Doing CSR Activities

4.1 ITC

In order to create new competitive advantages for its businesses as well as boosting natural capital and the livelihood of people, it has created fresh business models that are capable of delivering larger and immortal values.

4.1.1 Key Initiatives

Eradication of hunger and poverty includes crop productivity improvement, livestock/dairy development, and creation of sustainable livelihoods for the poor. Also, there are initiatives such as the promotion of education, gender equality & empowering women, promotion of women-based micro-enterprises, reducing child mortality & improving maternal health, basic hygiene and sanitation improvement initiatives combating diseases, etc.

4.2 Tata Power

The corporate values and operations of Tata Power are however to be integrated so as to fulfill the expectations of all its stakeholders including contractors as well as communities. The CSR vision and goal of the company are also provided along with its five main thrust areas for attention and geographical focus. The availability of funding or participation in projects beyond national boundaries will ensure that CSR programs are synchronized across locations. This policy is designed to guide resource, initiative development, execution, and assessment procedures.

4.2.1 Key Initiatives

Tata group had taken five total initiatives which include education, health and sanitation, livelihood & skill building, financial inclusivity, and water (drinking and irrigation). These are some of the performance indicators the organization created to self-evaluate themselves. Some of the indicators are; percentage increase in academic performance in English, Mathematics, and Science then boy girl ratio, enrolment, dropout percentage, etc. Others include policies (*yojana*) like *E-Vidya, Vidyasagar, Arogya, Samriddhi, Swajal*, and many more.

4.3 Infosys

The corporate social responsibility (CSR) activities of Infosys Limited were implemented early on. Encouraging social and environmental responsibility is just as important to comprehensive business growth as consistent economic performance. The company created the Infosys Foundation, and in order to fulfill its CSR obligations, it may create new entities or collaborate with existing ones ("CSR Implementing Agencies"). The company has always prioritized making the world a more liveable place for coming generations as well as supporting the sustainable growth of society and the environment.

4.3.1 Key Initiatives

Infosys mainly focused on Hunger, Poverty, Malnutrition and Health, Education, Rural Development Projects, Gender Equality and Empowerment of Women, Environmental Sustainability, National Heritage, Art and Culture, Disaster Management, etc. (Infosys, 2021).

4.4 Reliance

Speaking of CSR, Reliance Industries Limited (RIL)—one of the largest Indian conglomerates—has a rather structured and efficient CSR program active. The CSR of RIL encompasses provision of solutions in spheres of sustainable development, increase in the quality of population's life, and enhancement of socioeconomic development of regions where RIL operates. These programs have been set out, as a result of the vision of the Founder of this company Dhirubhai Ambani.

The clearest example of how large companies may also invest greatly in the improvement of the general welfare through properly developed and effective CSR initiatives is Reliance Industries Limited. RIL significantly boosts the performance of the socioeconomic pillar in India by focusing on several areas such as culture, sports, education, health, rural development, environment, and disaster response. By highlighting the corporations' role in promoting corporate social responsibility on the following fours aspects of sustainability, one can understand how the big companies can be the change makers for facilitating sustainable development.

4.4.1 Key Initiatives

Since RIL is a diverse firm, there are many ways that this firm can significantly increase and extend the social impact it has to offer while at the same time generating value for everyone. RIL's idea is to align the CSR programs in a way that the company focuses on particular social concerns and gives people an ability to sustain themselves. Thus, in the framework of such initiatives, the company has identified six priorities that it considers crucial for its social activity in order to act comprehensively in the sphere, thus defining its CSR strategy. There are some areas where RIL focused on such as Rural Transformation, Healthcare, Education, Environment Protection of National Heritage, Art and Culture, and Disaster Response.

4.5 Tata Steel

Over the course of its existence that is one hundred years, Tata Steel has worked hard for the betterment of Jharkhand in the socioeconomic terms. The Corporate Accountability Certification (CAC) program and strategy witnessed an improvement in the standard of living in specific communities in which the firm operates as the CSR committee's agenda item. They also initiated the "Lifeline Express" in 1991 which is actually a working train with a hospital on wheels to extend relief in the needy parts of India. They have conducted several awareness crusades on Acquired Immune Deficiency Syndrome or AIDS.

4.5.1 Key Initiatives

Some of Tata Steel's key initiatives are eradicating hunger, poverty, and malnutrition, promoting health care including preventive Healthcare, making available safe drinking water, sanitation, promotion of education including special education, reducing inequalities faced by socially and economically backward groups, environmental sustainability, etc (Fig. 1).

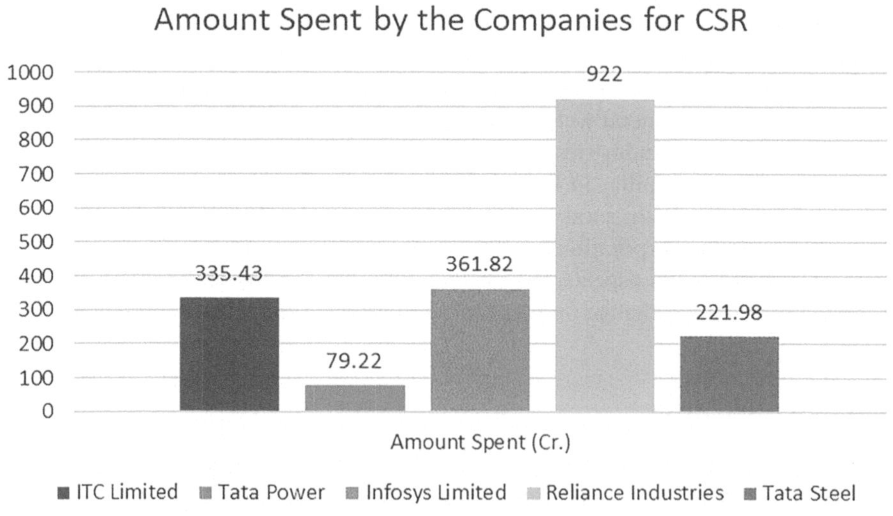

Fig. 1 Chart depicting the amount spent by the companies for CSR (INR Cr)

5 The Benefits to Private Firms and the Government

Business sustainability, also denoting corporation's responsibility to generate positive impact to the society, has become an essential aspect of modern-day business ventures and corporations. Since CSR is more of a voluntary activity, there is need for governments to intercede and encourage/ support these programs. This chapter discusses the several roles that governments have adopted toward the implementation of CSR and focuses on the benefits of such initiatives to both sides, the government and the private entities. Governments and commercial businesses alike can gain much from CSR initiatives, some of those are listed below.

5.1 Benefits for Private Companies

- Better public perception: CSR enhances organizational image and perception in the eyes of the consumers hence increases loyalty and trust.
- Social license to operate: There is authority and certification given to corporations which engage in corporate social responsibilities for CSR.
- Employee satisfaction: Due to the fact that many people consider the social responsibility of organizations a key factor to join a certain company, CSR activities act as a talent magnet and productivity booster.
- Risk management: Managing social and environmental risks helps the businesses to minimize their legal risks, regulatory penalties, and damages to company image.
- Sustainable business practices: Business Sustainable Development Construct—Corporate Social Responsibility (CSR) increases profitability within the long term by providing a guideline for appropriate business procedures that enhance business performance.

5.2 Benefits to Government

- Social well-being: More specifically, Corporate Social Responsibility (CSR) relieves pressure on public assets since it tackles social issues involving, for instance, poverty, health, and education.
- Economic growth: Second, successful CSR can lead to the creation of a successful society and therefore be beneficial for the economy as a whole since the key to successful economy is first of all its stability.
- Public–private relationships: By fostering CSR, those governments can build the favorable partnerships with private sectors for the synergetic solutions of the social issues.

- Enhanced governance: While establishing and implementing CSR projects, the governments can reduce corruption level and enhance general governance.
- Environmental protection: As per the governmental goals for sustainable development and combating with climate change, environmental protection is commonly included in CSR programs.

6 Conclusion

CSR government policies play a large role in bearing social responsibility in countries such as the UK where firms donate over £1.5 billion a year to social and community projects. Thus, solving such issues as poverty, healthcare, or education, this investment also reduces the pressure on spending the taxpayers' money. Moreover, CSI leads to a more favorable perception of the company by the public, thus increasing public confidence, which enhances the morale of the employees as well as the company's sales. Besides, they ensure compliance with the laid down standards and disclose any injustice that brings down corruption levels and in turn foster structural-economic development and social order. We can conclude that both businesses and society are benefiting from Corporate Social Responsibility (CSR) activities. Taking action that can lead to the case of occupying less energy is an example of CSR that in one way or another lead to cost savings. An organization, which on the average spends \$500,000 a year on energy costs, will save \$100,000 in a year by reducing energy consumption level to 20%. Also, CSR enhances the image of a business and this can lead to the establishment of a large market share. Based on an example, a company that spend \$200,000 in the community may get 5% increase of the total sales due to enhanced brand reputation; if the company sells \$20 million per year, additional million dollars of sales comes to the company's benefit.

CSR rules are not as cheap and may also not be easy to put into practice, the aspect of green washing and poor governance may be observed in developing countries. This is because the Government of India mandates that companies should spend 2% of their net profit on CSR and many have criticized the above fact. This is much needed with respect to capital, especially in small and medium enterprises. Besides that, as many as 59% of the American consumers believed that companies engage in "greenwashing," which has a negative impact on the genuine CSR programs. Nevertheless, the governments persist in having a great challenge in ensuring the real business commitment and in maintaining an appropriate CSR drive against the practical hard facts. Initial costs are often coupled with CSR initiatives, which in turn may be burdensome for SMEs; For instance, a small business may spend up to \$50,000 to acquire environmentally sound equipment; this entails a huge cost that could compromise the seller's ability to feed himself. Second, while those organizations that operate with the primary aim of making a quick buck may be put off by CSR because of the likely long-time horizon for returns on investment. Hence, if a business embarks on spending \$300,000 on CSR initiatives, solving the country's problems for years then waiting for improvement in the profits may take several years then make the justification of the investment from a strategic management perspective a tough one.

References

Blundo, D. S., Sánchez, R. G., Salgado, S. M., & Muiña, F. E. G. (2021). Flexibility and resilience in corporate decision making: A new sustainability-based risk management system in uncertain times. *Global Journal of Flexible Systems Management, 22*, 107–132.

Bowen, H. R. (1953). *Social responsibilities of the businessman*. University Of Iowa Press.

Branco, M. C., & Rodrigues, L. L. (2006). Corporate social responsibility and resource-based perspectives. *Journal of Business Ethics, 69*, 111–132.

Conway, E. (2018). Sustainability, the triple bottom line and corporate social responsibility. *Contemporary Issues in Accounting*, 15–35. https://doi.org/10.1007/978-3-319-91113-7_2

Harvey, C., Maclean, M., Gordon, J., & Shaw, E. (2011). Andrew Carnegie and the foundations of contemporary entrepreneurial philanthropy. *Business History, 53*, 425–450. https://doi.org/10.1080/00076791.2011.565516

IBM. (2021). What is sustainability in business? | IBM. www.ibm.com. https://www.ibm.com/topics/business-sustainability

Iglesias, O., Markovic, S., Bagherzadeh, M., & Singh, J. J. (2020). Co-creation: A key link between corporate social responsibility, customer trust, and customer loyalty. *Journal of Business Ethics, 163*, 151–166. https://doi.org/10.1007/s10551-018-4015-y

Infosys. (2021). *Corporate social responsibility policy*. INFOSYS LIMITED.

Kitzmueller, M., & Shimshack, J. (2012). Economic perspectives on corporate social responsibility. *Journal of Economic Literature, 50*, 51–84.

Maloni, M. J., & Brown, M. E. (2006). Corporate social responsibility in the supply chain: An application in the food industry. *Journal of Business Ethics, 68*, 35–52. https://doi.org/10.1007/s10551-006-9038-0

McBarnet, D. (2009). Corporate social responsibility beyond law, through law, for law. *SSRN Electronic Journal*. https://doi.org/10.2139/ssrn.1369305

Nie, P., Wang, C., & Meng, Y. (2019). An analysis of environmental corporate social responsibility. *Managerial and Decision Economics, 40*, 384–393. https://doi.org/10.1002/mde.3009

O'brien, J. (2005). *Governing the corporation: Regulation and corporate governance in an age of scandal and global markets*. Wiley.

Smith, A. D. (2007). Making the case for the competitive advantage of corporate social responsibility. *Business Strategy Series, 8*, 186–195. https://doi.org/10.1108/17515630710684187

Windsor, D. (2001). The future of corporate social responsibility. *The International Journal of Organizational Analysis, 9*, 225–256. https://doi.org/10.1108/eb028934

The Green Revolution in Finance: Bonding with the Planet Through Green Bonds

Megha Rani Patel, Bateshwar Singh, and Ajay Pratap Yadav

Abstract Green bonds, with a total issuance of $3.143 trillion, have become essential instruments for funding sustainable development and combating climate change. This chapter analyzes their exponential growth, including $328.1 billion issued in 2024, and examines regional trends, highlighting Europe's leadership ($1383.3 billion) and Asia-Pacific's rapid growth (61.25%). Case studies, such as Yes Bank's green bonds financing 3.61 GW of renewable energy projects, underscore tangible benefits like CO2 emission reductions. The integration of green bonds into global stock exchanges enhances their visibility, liquidity, and adoption. This analysis establishes green bonds as vital financial tools for advancing a low-carbon economy and fostering global sustainability.

1 Introduction

Green bonds are debt securities dedicated to funding projects aimed to deal with climate change. These bonds are primarily issued in various forms including self-labeled corporate bonds, green asset-backed securities, green project bonds, and bonds from supranational, international, governmental, or municipal entities.

Green bonds have become essential financial instruments dedicated to supporting projects that benefit the environment, significantly aiding the global shift toward resource-efficient and low-carbon economies (Chen & Zhao, 2021). The International Capital Market Association (ICMA) defines a green bond as a bond instrument where proceeds are specifically directed to finance or refinance, in whole or in part, new and/or existing eligible Green Projects. This definition highlights the primary aim of green bonds: funding projects with environmental benefits, setting them apart from traditional bonds by their specific use of proceeds.

M. R. Patel (✉) · B. Singh · A. P. Yadav
Department of Commerce and Financial Studies, Central University of Jharkhand, Ranchi, India
e-mail: megha.21330202003@cuj.ac.in; bateshwar.singh@cuj.ac.in; ajay.pratap@cuj.ac.in

A. Sharma et al. (eds.), *The Synergistic Effect of Sustainable Business Practices on Corporate Performance*, https://doi.org/10.1007/978-981-96-3375-3_9

The growing issuance and market demand for green bonds have led to the establishment of clear standards and criteria for identifying eligible green projects (Climate Bond Initiative). These bonds are structured similarly to conventional bonds, ensuring repayment of the borrowed amount over a specified period, along with fixed or variable interest payments to creditors. However, the distinguishing factor is their exclusive purpose: financing or refinancing green projects (Clapp & Pillay, 2017).

Green bonds have gained traction among various entities, including development banks, government bodies, and private companies, as a means to generate capital for green investments. These investments are crucial for addressing pressing environmental issues such as global warming, energy conservation, pollution control, and overall environmental protection (Agliardi & Agliardi, 2019). According to Rao (2021), the financial market's increasing focus on "Green Finance"—encompassing green bonds, Environmental, Social, and Governance (ESG) criteria, and renewable energy sources—reflects a broader commitment to funding initiatives that protect the environment.

In emerging economies like India, green bonds offer a promising way to attract investments aimed at climate action. Transitioning to a Low-Carbon Climate-Resilient (LCR) future requires significant investments, particularly in a country that relies heavily on affordable fossil fuels and faces considerable price volatility (Chen & Zhao, 2021). Green bonds provide a mechanism to finance the necessary investments for reducing greenhouse gas emissions and promoting sustainable development (Kung et al., 2022).

The urgency of addressing climate change and its associated economic and social impacts is increasingly clear. Natural disasters, intensified by climate change, have caused substantial economic losses globally, with developing countries being the most affected. For instance, the 2018 floods in Kerala (United Nations Development Programme, 2018) and the 2020 cyclone Amphan in West Bengal and Odisha (World Meteorological Organization, 2021) resulted in significant financial damages and displaced millions of people. The economic cost related to climate disasters underscores the critical need for sustainable financing mechanisms (Tolliver et al., 2019).

By channeling capital toward environmentally friendly projects such as renewable energy, energy efficiency, and clean transportation, green bonds play a crucial role in combating climate change (Sisodia et al., 2022). The issuance of green bonds has grown remarkably, with developed countries like the United States, France, and China leading the way. However, the potential for green bonds in developing countries remains substantial, as these nations strive to balance economic growth with environmental sustainability (Sartzetakis, 2021).

In summary, green bonds are essential for financing the transition to a sustainable, low-carbon economy (Cheong & Choi, 2020). They provide the necessary capital for green projects, addressing both environmental challenges and fostering economic growth (Bohringer et al., 2015). As the world faces the urgent need to mitigate climate change, the importance of green bonds in facilitating this transition cannot be overstated (Fatica & Panzica, 2021).

2 Differences Between Green Bonds and Traditional Bonds

According to Zhou et al. (2024), understanding the difference between green bonds and conventional/traditional bonds is crucial for investors and issuers. Green bonds are specifically designed to fund environmentally friendly projects (Li & Jia, 2017), whereas traditional bonds are utilized for a wide range of purposes. Table 1 explores key differences in purpose, transparency, market appeal, flexibility, regulation, risk profile, and pricing.

3 Historical Development and Evolution

Monk and Perkins (2020) segmented the development of the green bond market into five distinct phases, i.e. creation and initial issuance from 2006 to 2009 where the European Investment Bank (EIB), which launched its first Climate Awareness Bond (CAB) in 2007 followed by another multilateral development bank the World Bank

Table 1 Difference between green bonds and traditional bonds

Aspect	Green bonds	Traditional bonds
Purpose of investment	Proceeds raised from issue of green bonds are used for environmentally friendly and climate change related projects	Proceeds raised from issue of traditional bonds are used to for a wide range of purposes without any specific constraints/themes.
Transparency and reporting	Detailed documentation regarding the environmental benefits of the funded projects is essential. Additionally, to establish their validity, green bonds need certifications from independent second parties before issuance and ongoing reports after issuance.	Generally, no special reporting requirements on the use of proceeds are necessary.
Market appeal	Attract environmentally conscious investors and can benefit from green premiums in some markets (Flammer, 2021)	Attract to a broad range of investors, driven by credit quality and yield.
Flexibility	Less flexible	More flexible
Regulation and certification	Green bonds necessitate verifications from independent third parties in the market to maintain credibility, which in turn leads to increased costs related to compliance and issuance.	No specific environmental certifications are required.
Risk profile	May carry additional reputational risk if the projects do not deliver the expected environmental benefits.	Risk is generally associated with the issuer's creditworthiness and market conditions.
Pricing	Can potentially offer a lower yield compared to traditional bonds due to higher demand from specific investor groups.	Yield is primarily determined by the credit risk, duration, and market conditions.

issued its first bond officially labeled as a "green bond" in November 2008. In the second stage, there was an initial development and the International Finance Corporation's (IFC) $1 billion benchmark issuance in March 2013 and market started growing. In the third stage, the state of Massachusetts issued the first municipal bond in July 2013, with the city of Gothenburg following in October. November saw the corporate issuances started happening, with three releases in 3 days, including a record €1.4 billion offering by Électricité de France (EDF). Around this time, Zurich Insurance committed $1 billion to green bond investments, the first in a series of pledges from major institutional investors. A significant development took place in January 2014 with the introduction of the Green Bond Principles (GBPs), which are voluntary guidelines aimed at enhancing transparency, disclosure, and market integrity (ICMA, 2022). As an impact, a major take-off was observed in green bond issuance during 2014 and 2015. During the same period, the green bond market reached $ 100 billion (de Chanrond, 2023). In the final stage as divided by Monk and Perkins (2020) as the stage of diversification in the green bond issuances by issuers coming from the emerging markets, starting with South Korea and Taiwan. China led in 2016, with notable issuances from Brazil, India, and Poland.

To date, cumulative green bond issuance has reached $3.143 trillion. In 2024 alone, $328.1 billion has been issued, adhering to the Climate Bonds Initiative (CBI) definitions. This includes $19.3 billion in certified climate bonds and $308.8 billion in labeled bonds that align with CBI's "Green" definitions. Additionally, $35.8 billion in labeled green bonds that do not meet CBI definitions have been excluded from the 2022 issuance totals. Monthly issuance for 2024 is as follows: $81.274 billion in January, $54.381 billion in February, $67.358 billion in March, $42.564 billion in April, $76.153 billion in May, and $6.342 billion in June.

4 Trends Over the Years

Figure 1 illustrates the bond issuance trends by region from 2014 to 2023 which reveals significant insights into regional market dynamics. Europe has experienced the most notable growth in bond issuance, with a sharp increase beginning in 2017 and peaking in 2023. Asia-Pacific also shows substantial growth, particularly from 2016 onward, reaching its highest level in 2023. North America displays a steady increase but with a slight decline in 2023. Supranational issuance remained relatively flat until a notable rise in 2021, followed by a dip in 2023. Both Africa and Latin America exhibit much smaller issuance volumes, with slight fluctuations but generally low compared to other regions.

To understand these trends more comprehensively, we analyzed the total bond issuance, compound annual growth rate (CAGR), and volatility for each region. Europe leads with the highest total bond issuance of 1383.3 billion, followed by Asia-Pacific with 727.9 billion, and North America with 513.0 billion. Supranational issuers recorded 91.0 billion, Latin America 49.3 billion, and Africa the lowest at 8.1 billion. Asia-Pacific exhibits the highest growth rate with a CAGR of 61.25%,

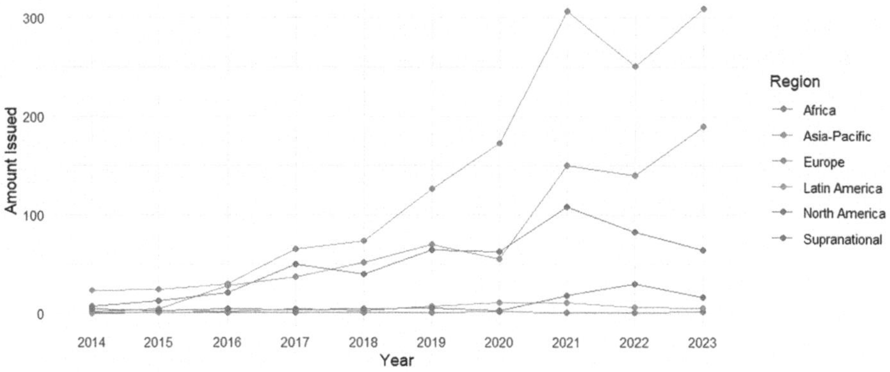

Fig. 1 Regional trends over the years. Source: Authors compilation

indicating rapid market expansion. Latin America follows with a CAGR of 37.69%, showing significant growth despite a smaller total issuance. Europe's CAGR stands at 29.47%, North America at 24.17%, Africa at 17.46%, and Supranational issuers at the lowest with 15.22%.

Volatility, measured by standard deviation, highlights Europe as the most volatile region with 114.98 billion, reflecting large annual variations in issuance. Asia-Pacific's volatility is 65.04 billion, North America's 31.87 billion, Supranational issuers 9.13 billion, Latin America 3.74 billion, and Africa the least volatile at 0.55 billion, due to consistently low issuance.

Europe's leadership in total issuance and considerable market volatility, suggesting a dynamic environment with large fluctuations. Asia-Pacific's substantial total issuance and the highest growth rate indicate a rapidly expanding market. In contrast, Supranational and African markets, while smaller and less volatile, exhibit steady growth rates, suggesting potential stability and consistent expansion. Recent years have highlighted significant growth in the Asia-Pacific and Supranational sectors of the green bond market, while Africa and Latin America have shown relatively low new issuances.

In Fig. 2, the analysis of bond issuance trends from 2014 to 2023 across Developed, Emerging, and Supranational markets reveals distinct patterns and growth. Developed markets have experienced the most significant and steady growth, with a particularly sharp surge in bond issuance from 2020 to 2021, indicating robust market activity and substantial financial movements. Emerging markets, on the other hand, show a gradual but consistent rise in bond issuance, highlighting a strengthening of financial activities and increasing engagement with international capital markets. Supranational issuers, although starting from a lower base, demonstrate modest growth up until 2017, followed by a noticeable increase from 2021 onward, reflecting growing funding requirements for supranational initiatives. The total bond issuance data further underscores these trends, with Developed markets leading at 1966.4 billion, Emerging markets at 617.9 billion, and Supranational issuers at 188.4 billion. The Compound Annual Growth Rate (CAGR) analysis

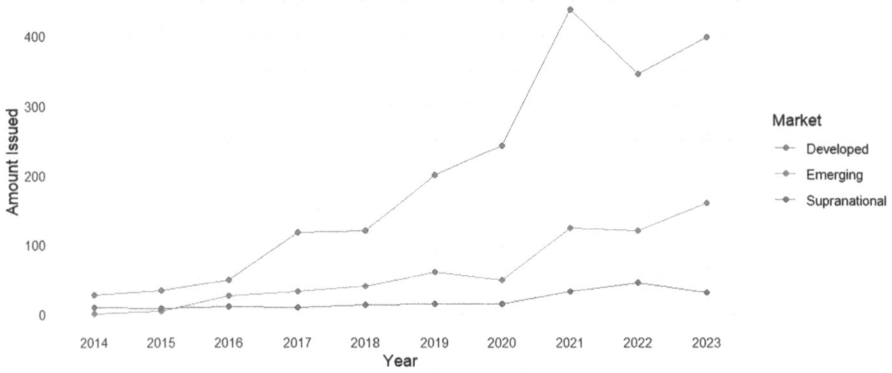

Fig. 2 Market trends over years. Source: Authors compilation

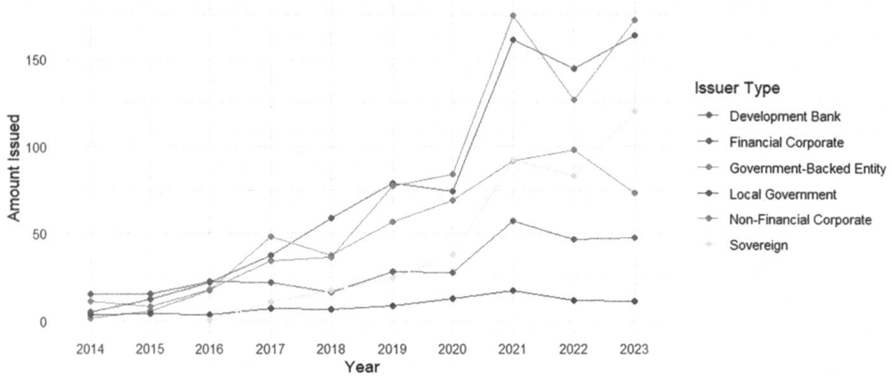

Fig. 3 Amount issued by different types of issuers over time. Source: Authors compilation

shows Emerging markets with the highest growth rate at 69.82%, demonstrating rapid expansion and attracting increasing investments. Developed markets follow with a CAGR of 30.93%, and Supranational entities show the lowest growth at 12.78%. Volatility measures reveal that Developed markets have the highest variability, indicating dynamic and possibly reactive market conditions, while Emerging markets exhibit considerable volatility and Supranational markets show the least volatility, suggesting more stable issuance patterns.

Figure 3 presents the trends in bond issuance from 2014 to 2023, segmented by various types of issuers. Each line represents a different issuer type, depicting how their bond issuance volumes have evolved over the years. Across the board, there is a general upward trend in bond issuance, indicating a growing reliance on or need for bond financing across different sectors. Development Banks show a consistent rise in bond issuance until 2021, after which there is a slight decline, followed by stabilization. Financial Corporates experience a significant surge, particularly from 2020 onward, peaking in 2023. This could reflect robust market activity and possibly an increased need for capital due to economic shifts. Government-Backed

Entities also display a noticeable increase, with a sharp rise in 2021 and 2022, before a minor decrease in 2023, suggesting periods of heightened financing needs.

Local Governments show a more gradual and steady growth in bond issuance, indicating a consistent but moderate need for additional funding over the decade. Non-Financial Corporates demonstrate rapid growth from 2018, peaking in 2021, which then slightly tapers off but remains high, highlighting substantial financing demands in this sector. Sovereign issuers, starting from almost negligible amounts in 2016, see rapid increases, particularly notable in 2021, suggesting significant governmental financing needs during this period, potentially due to responses to global events like the pandemic.

Overall, a dynamic bond market with increasing activity over the years, underlining how different entities have navigated changing economic conditions and financing environments. The sharp increases around 2020 and 2021 across several issuer types likely reflect responses to economic disruptions caused by global events such as the COVID-19 pandemic, underscoring the bond market's role in economic recovery and stabilization efforts.

Figure 4 shows the funding allocation trends across various key sectors from 2014 to 2023, indicating where bond issuance funds have been directed over these years. Each line represents a different sector and shows the fluctuation in the amounts issued annually. The sectors of Buildings and Energy exhibit significant variations, with the Energy sector peaking sharply in 2021, suggesting a substantial increase in funding, possibly due to heightened focus on energy initiatives or large-scale projects during that period. The buildings sector mirrors this peak, which could be attributed to major construction or infrastructure developments.

ICT and Transport sectors demonstrate a steadier, gradual increase in funding over the years, reflecting ongoing investment in technological infrastructure and transportation facilities, respectively. The rise in these sectors is consistent, peaking in 2021, hinting at strategic expansions or enhancements.

Other sectors such as Industry, Land Use, Unspecified A&R, Waste, and Water show relatively steady and modest growth throughout the period. This consistent

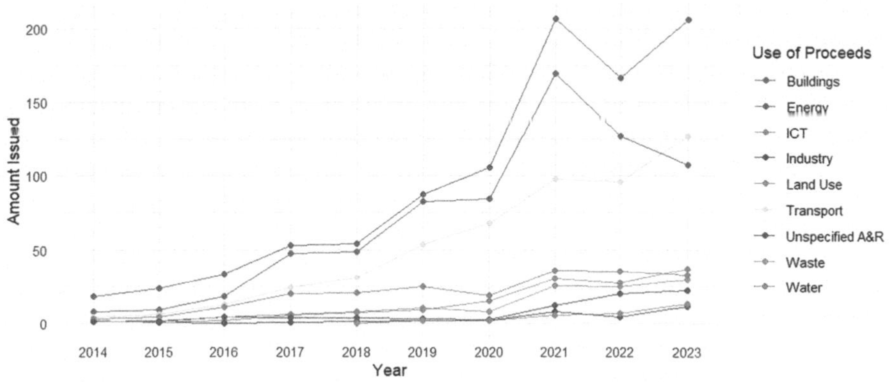

Fig. 4 Funding allocation: Key sectors where issued amounts have been utilized. Source: Authors compilation

Table 2 Types of green bonds

Type	Proceeds usage	Debt recourse	Example
Use of proceeds bond	Allocated to green projects	Issuer's credit rating applies to the bond	EIB Climate Awareness Bond; Barclays Green Bond
Use of proceeds revenue bond or ABS	Allocated to or refinances green projects	Revenue streams from fees, taxes, etc., collateralize the debt	Hawaii State (collateralized by fee on electricity bills)
Project bond	Dedicated to specific green project(s)	Recourse is solely to the project's assets and balance sheet	Invenergy Wind Farm (backed by Invenergy Campo Palomas wind farm)
Securitization (Assest Backed Securities) bond	Refinances green project portfolios or earmarked for green projects	Recourse to a group of projects pooled together	Tesla Energy (backed by residential solar leases); Obvion (backed by green mortgages)
Covered bond	Allocated to eligible projects in the covered pool	Recourse to the issuer and, if issuer defaults, to the covered pool	Berlin Hyp green Pfandbrief; Sparebank 1 Bolligkredit green covered bond
Loan	Allocated to eligible projects or secured on eligible assets	Full recourse to borrower(s) for unsecured loans; recourse to collateral for secured loans, may have limited recourse to borrower(s)	MEP Werke; Ivanhoe Cambridge and Natixis Assurances (DUO); OVG
Other debt instruments	Allocated to eligible projects		Convertible Bonds, Notes, Schuldschein, Commercial Paper, Sukuk, Debentures

Source: Climate Bond Initiative

investment indicates ongoing but less dramatic financing needs compared to the more volatile sectors like Energy or Buildings.

A significant observation is the sharp decline in funding for the Energy sector post-2021, which might suggest the completion of significant projects or a shift in investment priorities away from this sector. Overall, Fig. 4 shows how funding priorities have shifted over the years, reflecting strategic investment decisions which might be influenced by varying factors such as market demands, policy changes, or broader economic conditions. Table 2 exhibitis dedicated sections for green and sustainable bonds.

Name of Stock Exchange	Type of Dedicated Section	Launch Date
Oslo Stock Exchange	Green bonds	January 2015
Stockholm Stock Exchange	Sustainable bonds	June 2015
London Stock Exchange	Sustainable bonds	July 2015
Shanghai Stock Exchange	Green bonds	March 2016
Mexico Stock Exchange	Green bonds	August 2016
Luxembourg Stock Exchange	Luxembourg Green Exchange	September 2016
Borsa Italiana	Green and Social bonds	March 2017
Taipei Exchange	Sustainable bonds	May 2017
Johannesburg Stock Exchange	Green bonds	October 2017
Japan Exchange Group	Green and Social bonds	January 2018
Vienna Exchange	Green and Social bonds	March 2018
Nasdaq (Multiple stock exchanges)	Sustainable bonds	May 2018
Swiss Stock Exchange	Green and Sustainability bonds	July 2018
The International Stock Exchange	Sustainable bonds	November 2018
Frankfurt Stock Exchange	Green bonds	November 2018
Santiago Stock Exchange	Green and social bonds	July 2019
Euronext (Multiple stock exchanges)	Green, Sustainable and Social bonds	November 2019
Bombay Stock Exchange	Green bonds	June 2019
Argentina Stock Exchange BYMA	Green, Sustainable and Social bonds	September 2019
Brazil Stock Exchange	Green bonds	September 2019
Nigerian Stock Exchange	Sustainable bonds	October 2019
Hong Kong Exchange	Green, Sustainable and Social bonds	June 2020
Korea Exchange	Green, Sustainable and Social bonds	June 2020
Toronto Stock Exchange	Sustainable bonds	November 2020
Singapore Stock Exchange	Green, Sustainable and Social bonds	n.a
Bolsas y Mercados Españoles (BME)	Green, Sustainable and Social bonds	n.a

Fig. 5 Global stock exchanges and their dedicated sections for green and sustainable bonds: A timeline overview. Source: Climate Bond Initiative

5 Types of Green Bonds and Dedicated Section in Stock Exchange for the Green Bonds

Figure 5 shows how the global stock exchanges have increasingly recognized the importance of sustainable finance, leading to the creation of dedicated sections for green, sustainable, and social bonds. This initiative began with the Oslo Stock Exchange, which launched its green bonds section in January 2015. This was followed by the Stockholm Stock Exchange which launched a dedicated section for sustainable bonds in June 2015. The London Stock Exchange, Shanghai Stock Exchange and Mexico Stock Exchange did the same in July 2015, March 2016 and August 2016 respectively (Alam and Akhtar (2024).

Luxembourg's Stock Exchange took a significant step by launching the Luxembourg Green Exchange in September 2016, providing a dedicated platform for green bonds. Italy's Borsa Italiana followed in March 2017 with a section for green and social bonds, while the Taipei Exchange introduced a sustainable bonds section in May 2017. In October 2017, the Johannesburg Stock Exchange also launched a green bonds section.

The Japan Exchange Group and the Vienna Exchange both introduced sections for green and social bonds in January and March 2018, respectively. In May 2018, Nasdaq's multiple stock exchanges started offering a sustainable bonds section. The Swiss Stock Exchange launched a section for green and sustainability bonds in July 2018, followed by the International Stock Exchange with a sustainable bonds section in November 2018, and the Frankfurt Stock Exchange with a green bonds section the same month.

The Santiago Stock Exchange introduced a section for green and social bonds in July 2019, and Euronext, covering multiple stock exchanges, launched a section for green, sustainable, and social bonds in November 2019. Along with sections for sustainable and social bonds, the Bombay Stock Exchange (BSE) and the Argentina Stock Exchange BYMA both established sections for green bonds in June and September of 2019, respectively.

September 2019 saw the debut of a green bonds section on the Brazil Stock Exchange, while October 2019 saw the launch of a sustainable bonds section on the Nigerian Stock Exchange. Green, sustainable, and social bonds sectors were added to the Korea Exchange and the Hong Kong Exchange in June 2020. In November 2020, the Toronto Stock Exchange launched a segment dedicated to sustainable bonds. Although exact debut dates were not given, the Singapore Stock Exchange and the Bolsa y Mercados Españoles (BME) have also added sections for green, sustainable, and social bonds.

These initiatives across various global stock exchanges underscore the growing commitment to sustainability and the mobilization of capital toward environmentally and socially responsible projects. Table 3 exhibits the level of financial leverage used by various issuers to achieve renewable energy goals.

6 The Use of Financial Leverage to Achieve India's Renewable Energy Goals

Very High Leverage The Power Finance Corporation Ltd., a significant government-owned financial institution in India, has been a cornerstone in progressing the nation's power sector for over three decades. Committed to the Paris Agreement, India targets to derive about 40% of its electric power from non-fossil fuels by 2030, reducing its GDP emissions intensity by 33% to 35% from 2005 levels. In this vein, PFC actively promotes renewable energy projects, aiming to achieve an installed capacity of 175 GW by 2022, which will rise to 450 GW by 2030. As of March 2021, India's renewable capacity reached 94.3 GW, accounting for 22% of total power capacity, attracting global investors with the promise of low-risk, high-return, and tax-efficient opportunities.

Simultaneously, the Indian Railway Finance Corporation Ltd. under its Green Bond Framework is concentrating on funding sustainable transport infrastructure to support the Indian Railways' goal to increase its land-based freight transport,

Table 3 Leverage risk analysis for companies issuing green bonds

Sl. No.	Issuer name	Total assets	Total debt	Debt ratio	Total equity	Debt-to-equity ratio	Observations
1.	Adani Green Energy Ltd	673,610	542,240	80.50%	131,370	4.13	High leverage
2.	REC Ltd	4,655,030	3,810,000	81.85%	845,030	4.51	High leverage
3.	JSW Hydro Energy Ltd	487,417	250,510.7	51.40%	236,906.3	1.06	Moderate
4.	Indian Railway Finance Corp Ltd	4,910,000	4,190,000	85.34%	720,000	5.82	Very high
5.	Power Finance Corp Ltd	8,960,000	7,510,000	83.82%	1,450,000	5.18	Very high
6.	Mindspace Business Parks REIT	227,231	54,662	24.06%	172,569	0.32	Low
7.	Indian Renewable Energy Development Agency Ltd	367,084.03	92,291.39	25.14%	274,792.64	0.34	Low
8.	Axis Bank Ltd	13,440,000	2,160,000	16.07%	11,280,000	0.19	Very low
9.	Yes Bank Ltd	3,550,000	785,051.63	22.11%	2,764,948.37	0.28	Low
10.	Yarrow Infrastructure Pvt. Ltd	16,020	7357	45.92%	8663	0.85	Moderate
11.	Clean Sustainable Energy Pvt. Ltd	2474.33	1279.58	51.71%	1194.75	1.07	Moderate
12.	ReNew Pvt. Ltd	620,450	444,013	71.56%	176,437	2.52	Moderate leverage
13.	Adani Renewable Energy RJ Ltd./Kodangal Solar Parks Pvt. Ltd./ Wardha Solar Maharashtra	41,804	31,101	74.40%	10,703	2.91	Moderate leverage

Source: Authors calculation

contributing significantly to India's climate change efforts. Both PFC and IRFC exhibit high financial leverage, reflecting their strategic use of debt to fuel large-scale, environmentally sustainable projects. This leverage allows them to maximize the infusion of capital into critical infrastructure, enhancing both India's economic growth and environmental health.

High Leverage Rural Electrification Corporation Limited (REC) and Adani Green Energy Ltd. demonstrate significant financial leverage, strategically utilized to harness the burgeoning potential within the renewable energy sector. This leverage primarily arises from their bold investment strategies designed to enhance their renewable energy output, addressing India's increasing energy needs and environmental commitments.

REC, a key player in India's power infrastructure sector, invests heavily in a variety of renewable projects, including solar, wind, biomass, and small hydro projects. By leveraging debt financing, REC can embark on large-scale projects that align with both governmental renewable energy objectives and its Corporate Social Responsibility (CSR) initiatives. This approach not only boosts its project implementation capacity but also capitalizes on government support and the green bond market, thereby facilitating significant infrastructure investments with limited immediate equity use.

Adani Green Energy Ltd., on the other hand, exhibits high leverage due to its rapid expansion in solar and wind energy projects. The company employs debt financing, such as green bonds, to gather essential funds for its large-scale and capital-intensive renewable projects. This reliance on debt illustrates their strategic engagement with financial markets to support substantial growth and develop major renewable assets crucial for long-term environmental sustainability.

In essence, the high leverage for both companies is a strategic choice that supports their extensive investments in renewable energies, underscoring their dedication to reshaping India's energy landscape and advancing environmental and social well-being.

Moderate to Low Leverage Companies like JSW Hydro Energy Ltd., Yarrow Infrastructure Pvt. Ltd., and Clean Sustainable Energy Pvt. Ltd. have moderate leverage, which may reflect a balanced approach to funding their operations and projects. Adani Renewable Energy RJ Ltd./Kodangal Solar Parks Pvt. Ltd./Wardha Solar Maharashtra issuer falls into the moderate leverage category with a debt ratio of 74.40% and a debt-to-equity ratio of 2.91. Although the absolute asset and debt values are smaller compared to others, the ratio indicates a higher dependency on debt. This pattern suggests aggressive investment strategies in renewable energy projects, which is typical in the green energy sector due to high initial costs and long-term payoff structures. Whereas ReNew Pvt. Ltd. shows a debt ratio of 71.56% and a debt-to-equity ratio of 2.52, which places it in a category of moderate leverage. This indicates a substantial reliance on debt financing, albeit less so than some of the very highly leveraged entities in the dataset. Given ReNew Pvt. Ltd's focus on renewable energy, its debt may finance large-scale projects that have favorable long-term revenue potentials, albeit with a significant upfront capital requirement.

Low Leverage Mindspace Business Parks REIT, Indian Renewable Energy Development Agency Ltd., Axis Bank Ltd., and Yes Bank Ltd. show low leverage, indicating a conservative approach to debt, which could appeal to risk-averse investors, especially in sectors like real estate and banking related to sustainable projects.

7 Case Study on Yes Bank

India's first green bond was issued by Yes Bank in February 2015 and has been a pioneer in green financing within the country, initially raising INR 10 billion to support green projects. Following this milestone, the bank issued additional green bonds worth INR 3.15 billion in August 2015 and INR 3.3 billion in September 2016. In order to evaluate the effects of green bonds, the bank examined the projects that were supported with the profits, paying particular attention to environmental risk mitigation and the reduction of dangerous gases including carbon dioxide ($CO2$), sulphur dioxide ($SO2$), and nitrogen oxides (NOx).

Since its launch, the Indian green bond market has expanded dramatically, topping INR 50,000 crore through about 25 issuances, with more development anticipated. These bonds have drawn responsible investments from both domestic and global sources, providing India with much-needed funding for climate action. The growth of Yes Bank can be attributed in large part to its efforts, particularly to its two distinct green bond investments made in August 2015 through the Green Masala Bond program with the International Finance Corporation (IFC) and December 2016 through another green bond with the Dutch Development Bank, FMO. As of March 31, 2018, the bank had used the money from these green bonds to finance 2.73 GW of wind and solar energy projects.

Yes Bank was a key player in the market dynamic for green bonds in 2020, which grew despite global challenges. Up until March 31, 2020, the bank's contributions covered the funding of 3.61 GW of solar and wind energy projects. The expected environmental and social benefits of the three green bonds are assessed in this edition of YES BANK's green bond impact report. Over their lifetime, the electricity generation is projected to be 173.88 million MWh, avoiding 163.45 megatons of $CO2$, 1398.03 kilotons of $SO2$, and 420.80 kilotons of NOx emissions, as well as reducing the dependency on fossil fuels by 14.94 megatons.

By 2021, the total green bonds issued globally exceeded USD 552.7 billion, marking a 75% increase over 2020. Approximately 44% of all international bonds issued from India in 2021 were green, social, sustainable, or sustainability-linked bonds, with YES BANK continuing to lead in this space. Their green bond issuances have facilitated financing renewable energy projects worth INR 14,000 crore, contributing to cumulative capacities of around 8.5 GW.

Figure 6 "$CO2$ Emissions Avoided Over Time by Yes Bank" shows the annual and lifetime $CO2$ emissions avoided, measured in megatons, from 2016–2017 to 2021–2022. It reveals two distinct trends: annual emissions avoided, represented by the blue line, and lifetime emissions avoided, represented by the green line. The blue line, denoting annual $CO2$ emissions avoided each year, starts at 2.3 megatons in 2016–2017 and gradually increases to 9.166 megatons by 2021–2022. This

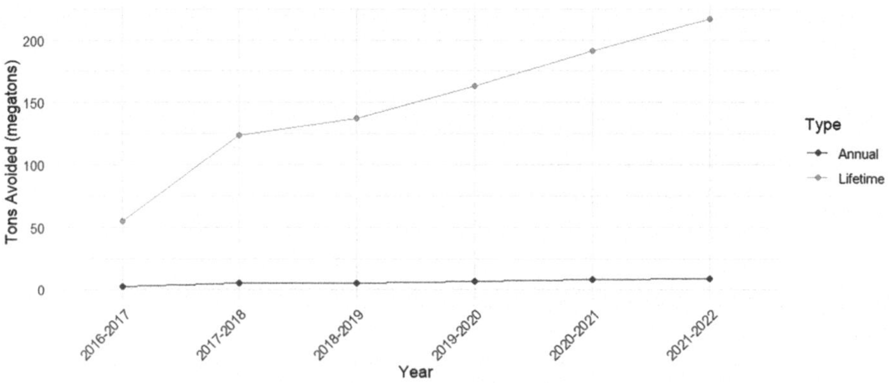

Fig. 6 CO2 emissions avoided over time by Yes Bank Source: Authors compilation

steady rise indicates a consistent effort by Yes Bank to mitigate CO2 emissions on an annual basis. In contrast, the green line, which represents cumulative or lifetime CO2 emissions avoided, begins at 55 megatons in 2016–2017 and shows a sharp increase each year, reaching 217.07 megatons by 2021–2022. The significantly steeper slope of the lifetime emissions line reflects the cumulative effect of ongoing projects and their substantial long-term impact on reducing CO2 emissions.

Figure 6 effectively highlights the growing impact of Yes Bank's green initiatives over time, with cumulative emissions avoided being substantially higher than annual emissions. This shows the long-term benefits of their green projects. The steady rise in both annual and lifetime CO2 emissions avoided underscores an ongoing and increasing commitment to reducing greenhouse gas emissions. The difference in scale between the annual and lifetime lines emphasizes how multiple years of green projects contribute significantly to reducing overall CO2 emissions in the long run. Overall, Yes Bank's contributions to mitigating climate change through their green financing efforts, with the increasing trend in both annual and lifetime CO2 emissions avoided underscoring the positive environmental impact of their investments in green projects over the years.

This case study was conducted using the Yes Bank's Green Bond Impact Reports published from the financial year 2016–17 to 2021–2022 which highlights Yes Bank's pioneering efforts and the transformative impact of green bonds on India's journey toward a low-carbon economy, emphasizing the crucial role of sustainable finance in addressing climate change challenges.

8 Conclusion

As discussed in this chapter, green bonds have emerged as crucial instruments for funding sustainable development (Chang et al., 2022) witnessing an exponential growth in issuance to \$3.143 trillion to date. This significant rise reflects their growing acceptance and effectiveness in mobilizing resources for environmentally beneficial projects (Bhutta et al., 2022). With $328.1 billion issued in 2024 alone, green bonds are not only reshaping investment landscapes but are also directly contributing to the global agenda against climate change.

The chapter discussed differing trends across various regions and issuer types highlights the versatility and global reach of green bonds. For instance, the Europe and Asia-Pacific regions have seen substantial growth in bond issuance, with Europe leading with a total issuance of $1383.3 billion and demonstrating a CAGR of 29.47%. Similarly, Asia-Pacific, with the highest growth rate at 61.25%, underscores the rapid expansion of green finance in emerging markets. Such data not only illustrate regional dynamics but also underscore the scale of financial commitment toward sustainable development.

Case studies, like the initiatives by Yes Bank, further concrete the real-world impact of these financial instruments. Yes Bank's strategic issuance of green bonds, culminating in the financing of 3.61 GW of wind and solar projects, exemplifies how targeted financial tools can foster substantial environmental and social benefits. The bank's success in reducing annual CO2 emissions, as detailed through their annual and lifetime CO2 emissions graphs, showcases the tangible outcomes of investing in green projects. Banga (2019) studied developing countries regarding the contribution of green bonds to the decrease of CO2 emissions.

Moreover, the broader adoption of green bonds by various global stock exchanges and their incorporation into mainstream financial markets is a testament to their increasing relevance. This trend not only enhances the visibility of green bonds but also ensures greater liquidity and market support for sustainable investment options, thereby encouraging more issuers to participate in green initiatives.

In summary, the journey of green bonds from a novel concept to a mainstream financial strategy underscores their potential to significantly influence global environmental policies and practices. As detailed through comprehensive market analysis, case studies, and specific data points, green bonds have proven their efficacy in channeling vast amounts of capital toward combating climate change. Looking forward, the continued evolution and acceptance of green bonds will be vital in financing the transition to a sustainable, low-carbon economy. They are not just financial instruments; they are pivotal components in the global strategy to foster a sustainable future, demonstrating that market-driven solutions can effectively address environmental challenges.

References

Agliardi, E., & Agliardi, R. (2019). Financing environmentally sustainable projects with green bonds. *Environment and Development Economics, 24*(6), 608–623. https://doi.org/10.1017/s1355770x19000020

Alam, M., & Akhtar, S. (2024). Introduction to green finance and technology integration. In *Green finance and technology: Integration and innovations (Chapter 1).* IGI Global. https://doi.org/10.4018/979-8-3693-1878-2.ch001

Banga, J. (2019). The green bond market: A potential source of climate finance for developing countries. *Journal of Sustainable Finance and Investment, 9*(1), 17–32. https://www.tandfonline.com/doi/full/10.1080/20430795.2018.1498617

Bhutta, U. S., Tariq, A., Farrukh, M., Raza, A., & Iqbal, M. K. (2022). Green bonds for sustainable development: Review of literature on development and impact of green bonds. *Technological Forecasting and Social Change, 175*, 121378. https://www.semanticscholar.org/paper/Green-bonds-for-sustainable-development%3A-Review-of-Bhutta-Tariq/6dc3093f3f62adbc005bcce912d801b530a425d4

Bohringer, C., Rutherford, T. F., & Springmann, M. (2015). Clean-development investments: An incentive-compatible CGE modeling framework. *Environmental and Resource Economics, 60*(4), 633–651. https://doi.org/10.1007/s10640-014-9762-3

Chang, L., Taghizadeh-Hesary, F., Chen, H., & Mohsin, M. (2022). Do green bonds have environmental benefits? *Energy Economics, 115*, 106356. https://www.sciencedirect.com/science/article/abs/pii/S0140988322004856

Chen, Y., & Zhao, Z. J. (2021). The rise of green bonds for sustainable finance: Global standards and issues with the expanding Chinese market. *Current Opinion in Environmental Sustainability, 52*, 54–57. https://doi.org/10.1016/j.cosust.2021.06.013

Cheong, C., & Choi, J. (2020). Green bonds: A survey. *Journal of Derivatives and Quantitative Studies: 선물연구, 28*(4), 175–189. https://doi.org/10.1108/JDQS-09-2020-0024

Clapp, C., & Pillay, K. (2017). Green bonds and climate finance. In *Climate finance: Theory and practice* (pp. 79–105). Climate Bond Initiative. https://www.climatebonds.net/market/explaining-green-bonds

de Chanrond, G. R. M. L. (2023). *Green bond regulatory frameworks subjecting finance to environmental objectives through "green" definitions*. Doctoral dissertation, The Chinese University of Hong Kong, Hong Kong.

Fatica, S., & Panzica, R. (2021). Green bonds as a tool against climate change? *Business Strategy and the Environment, 30*(5), 2688–2701. https://onlinelibrary.wiley.com/doi/10.1002/bse.2771

Flammer, C. (2021). Corporate green bonds. *Journal of Financial Economics, 142*(2), 499–516. https://www.sciencedirect.com/science/article/pii/S0304405X21000337

International Capital Market Association. (2022). Green bond principles: Voluntary process guidelines for issuing green bonds. https://www.icmagroup.org/assets/documents/Sustainable-finance/2022-updates/Green-Bond-Principles-June-2022-060623.pdf

Kung, C. C., Lan, X., Yang, Y., Kung, S. S., & Chang, M. S. (2022). Effects of green bonds on Taiwan's bioenergy development. *Energy, 238*, 121567. https://doi.org/10.1016/j.energy.2021.121567

Li, W., & Jia, Z. (2017). Carbon tax, emission trading, or the mixed policy: Which is the most effective strategy for climate change mitigation in China? *Mitigation and Adaptation Strategies for Global Change, 22*(6), 973–992. https://link.springer.com/article/10.1007/s11027-016-9710-3

Monk, A., & Perkins, R. (2020). What explains the emergence and diffusion of green bonds? *Energy Policy, 145*, 111641. https://www.sciencedirect.com/science/article/abs/pii/S030142152030375X

Rao, D. T. (2021). A study on green bonds in India–need of the hour. *IJRAR-International Journal of Research and Analytical Reviews (IJRAR), 8*(4), 62–69.

Sartzetakis, E. S. (2021). Green bonds as an instrument to finance low carbon transition. *Economic Change and Restructuring, 54*(3), 755–779. https://link.springer.com/article/10.1007/s10644-020-09266-9

Sisodia, G., Joseph, A., & Dominic, J. (2022). Whether corporate green bonds act as Armour during crises? Evidence from a natural experiment. *International Journal of Managerial Finance, 18*(4), 701–724. https://doi.org/10.1108/IJMF-10-2021-0501

Tolliver, C., Keeley, A. R., & Managi, S. (2019). Green bonds for the Paris agreement and sustainable development goals. *Environmental Research Letters, 14*(6), 064009. https://iopscience.iop.org/article/10.1088/1748-9326/ab1118

World Meteorological Organization. (2021). *State of the global climate 2021*. Retrieved from https://library.wmo.int/records/item/56300-state-of-the-global-climate-2021
Zhou, Y., Tian, L., Yang, X., & Wan, B. (2024). Robust green Schumpeterian endogenous growth model and spatial Kuznets curve. *Energy Economics, 133*, 107520. https://doi.org/10.1016/j.eneco.2024.107520

Websites

https://www.undp.org/publications/undp-annual-report-2018

Revolutionizing Circular Business Practices with Waste Reduction and Resource Optimization

Tien Thuy Tran Dang, Thinh Gia Hoang, Anna Maria Majo, Pallavi Pandey, and An Duong Thi Binh

Abstract The concept of the circular economy (CE) offers a transformative solution to pressing global challenges such as resource depletion, waste generation, and environmental degradation. By shifting from the traditional linear "take-make-dispose" model to a closed-loop system, the CE principles aim to enhance resource efficiency and sustainability. This paper proposes a Waste Reduction and Resource Optimization (WRRO)-oriented circular business model, offering a comprehensive theoretical framework to guide businesses in optimizing resource utilization and reducing waste. Through the exploration of academic literature and a practical case study, the paper illustrates how businesses can effectively transition to circular business models. It also addresses the challenges businesses face during implementation, providing actionable insights for both theory and practice. The paper's key contribution lies in integrating circular economy principles with resource efficiency strategies, offering a novel approach that supports both environmental sustainability and economic performance.

T. T. T. Dang · T. G. Hoang
International School of Business (ISB), UEH University, Ho Chi Minh City, Vietnam
e-mail: tien.dangtranthuy@westernsydney.edu.vn

A. M. Majo
Marian College, Kuttikkanam, India

P. Pandey
School of Commerce, Finance and Accountancy, Christ University, Pune, India

A. D. T. Binh (✉)
CIRTech Institute, HUTECH University, Ho Chi Minh City, Vietnam
e-mail: dtb.an@hutech.edu.vn

A. Sharma et al. (eds.), *The Synergistic Effect of Sustainable Business Practices on Corporate Performance*, https://doi.org/10.1007/978-981-96-3375-3_10

1 Introduction

The traditional linear economy, characterized by the "take-make-dispose" model, has long driven industrial production, focusing on short-term gains at the expense of environmental sustainability. This model, which relies heavily on the extraction of finite resources, has contributed to increasing environmental degradation, resource depletion, and waste accumulation (Romero-Hernández & Romero, 2018; Rathore & Sarmah, 2020). As industries and societies grapple with the environmental limits of this approach, there is an urgent need for more sustainable alternatives. In response, the circular economy (CE) has emerged as a transformative framework that seeks to decouple economic growth from resource consumption (Tanveer et al., 2024; Sheikh et al., 2024). CE principles advocate for a closed-loop system where waste is minimized, products are designed for longevity, and materials are continuously cycled back into the economy (Keijer et al., 2019). Despite the growing body of research on the benefits of CE practices—such as waste reduction, material reuse, and resource conservation—many businesses continue to face significant barriers in applying these principles effectively. Companies often struggle with resource flow optimization, the integration of circular principles into existing operations, and the economic viability of circular models (Tanveer et al., 2024; Serrano-Arévalo et al., 2024). Moreover, while circularity is often linked to environmental outcomes, the concept of resource efficiency—which involves maximizing the use of resources across all stages of production and consumption—remains underexplored in both academic literature and business practice (Desing et al., 2020; Romero-Hernández & Romero, 2018). This highlights the need for a more comprehensive framework that not only encourages waste reduction but also focuses on optimizing resource utilization within circular systems.

This chapter introduces the Waste Reduction and Resource Optimization (WRRO) framework, aimed at guiding businesses in their transition to circular business models. It critically examines how CE principles can not only reduce waste but also strategically optimize resource use across the entire product lifecycle. The chapter then applies this framework through a practical case study, illustrating how businesses can implement these concepts to drive sustainability and operational efficiency.

To illustrate the practical application of the proposed model, this chapter includes a case study that visualizes the dual-loop framework for resource flow. The case study features a global manufacturing company that has successfully integrated CE principles into its operations, focusing on optimizing resource utilization and minimizing waste. The company adopts a dual-loop CE model, combining both closed-loop (internal resource flow) and open-loop (external resource flow) systems. This approach facilitates a more dynamic and comprehensive model for resource optimization, ensuring that materials are continually recycled, remanufactured, or reused throughout the product lifecycle. The case study serves to demonstrate how the proposed model can be implemented in real-world business contexts, offering tangible examples of how companies can operationalize circular principles and improve resource efficiency while achieving sustainable growth.

This paper begins by introducing the concept of CE, highlighting its principles and the shift from linear to circular models focused on sustainability. It then explores how CE practices can advance waste reduction and resource efficiency, emphasizing the need for resource optimization. Building on this, the WRRO circular business model is proposed as a theoretical framework for businesses aiming to integrate these principles into their operations. A case study illustrates the application of the WRRO framework in practice, followed by an examination of the challenges businesses face when implementing circular models. The paper concludes with a discussion of the theoretical and practical contributions of the WRRO framework and offers future research suggestions to further refine CE practices and overcome implementation barriers.

2 The Concept of CE

The concept of CE has emerged as a transformative paradigm aimed at addressing the growing concerns over resource depletion, waste generation, and environmental degradation (Romero-Hernández & Romero, 2018; Rathore & Sarmah, 2020). At its core, the CE seeks to decouple economic growth from the consumption of finite natural resources by rethinking how we design, produce, and consume products (Keijer et al., 2019; Tanveer et al., 2024). The circular model is based on the principle of keeping products, materials, and resources in use for as long as possible, thereby creating closed-loop systems that regenerate natural systems. This approach is encapsulated through the framework of the 6Rs—Reduce, Reuse, Recycle, Recover, Redesign, and Remanufacture—which serve as guiding principles for businesses and policymakers seeking to transition toward more sustainable and efficient production systems (Tomić & Schneider, 2020; Chioatto & Sospiro, 2023). Each of these principles plays a crucial role in minimizing waste and enhancing resource efficiency. The "Reduce" principle emphasizes the need to decrease the number of raw materials used in production, while "Reuse" advocates for extending the lifespan of products by reusing them, either in their original form or as components (Rathore & Sarmah, 2020; Hidalgo et al., 2019). "Recycle" refers to the process of recovering valuable materials from products at the end of their life, whereas "Recover" focuses on reclaiming energy or materials from waste. "Redesign" involves creating products that are easier to repair, refurbish, or recycle, and "Remanufacture" entails rebuilding products or components to their original specifications, often using recycled or recovered materials (Tomić & Schneider, 2020; Prieto-Sandoval et al., 2019; Yang et al., 2024). These principles collectively aim to minimize waste, reduce the consumption of virgin resources, and create sustainable cycles of material use across industries.

The historical evolution of CE thinking can be traced back to traditional environmental sustainability models, which initially focused on minimizing environmental harm through strategies such as waste management and resource conservation (Gómez-Prado et al., 2022; Romero-Hernández & Romero, 2018). Early environmental movements emphasized reducing pollution, conserving biodiversity, and promoting sustainable agriculture and forestry practices. However, these approaches were often reactive, addressing the symptoms of unsustainable consumption rather than the root causes (Desing et al., 2020; Rathore & Sarmah, 2020). Over time, the limitations of the traditional linear "take-make-dispose" model became increasingly apparent, leading to the development of more proactive and holistic strategies (Prieto-Sandoval et al., 2019; Serrano-Arévalo et al., 2024). The contemporary CE framework, which emerged in the 1990s, built on the foundation of sustainability but introduced a more integrated and system-wide approach to resource management (Keijer et al., 2019; Tanveer et al., 2024). Scholars such as Kenneth Boulding, who in 1966 envisioned a "spaceship earth" economy that would cycle resources in closed loops, and later thinkers like William McDonough and Michael Braungart with their "Cradle to Cradle" design principles (Whalen & Whalen, 2018) began to shape the CE as a model that sought not only to reduce environmental impacts but to create value through the continual reuse and regeneration of resources (Baratsas et al., 2021; Hidalgo et al., 2019). The CE was thus envisioned not as a set of isolated environmental strategies, but as a comprehensive framework for transforming entire economic systems, wherein the design of products, business models, and consumer behavior all contribute to a sustainable future (Tanveer et al., 2024; Luthra et al., 2022).

In contrast to the linear economy, which is based on a model of extraction, production, consumption, and disposal, the CE is predicated on the idea of creating value without generating waste (Perey et al., 2018; Rathore & Sarmah, 2020). The linear model is inherently unsustainable as it relies on finite resources and produces waste that often ends up in landfills or oceans (Hidalgo et al., 2019; Najar et al., 2024). This model of take-make-dispose not only depletes natural resources but also contributes to environmental pollution and climate change. The CE, by contrast, promotes a regenerative system where the focus is on extending the lifecycle of products, closing the material loop, and decoupling growth from resource consumption (Chowdhury & Asiabanpour, 2024; Serrano-Arévalo et al., 2024). In practical terms, this means designing products that can be easily disassembled, reused, and recycled, establishing systems for product take-back and recycling, and fostering collaboration across supply chains to ensure that materials are continuously cycled back into production processes (Tanveer et al., 2024; Serrano-Arévalo et al., 2024). The circular model also emphasizes the creation of sustainable business models that generate economic value while reducing negative environmental impacts (Keijer et al., 2019). By shifting from a linear to a circular model, businesses can not only enhance their resource efficiency but also contribute to broader sustainability goals, reducing their ecological footprint and enabling the transition to a more sustainable, low-carbon economy (Prieto-Sandoval et al., 2019; Baratsas et al., 2021).

Thus, the shift from linear to circular models has significant implications for both resource efficiency and sustainability. It challenges the traditional economic assumptions of endless growth based on the extraction and consumption of finite resources, proposing instead a system that thrives on the regeneration of resources and the continuous cycling of materials (Romero-Hernández & Romero, 2018; Yang et al., 2024). The CE, with its focus on reducing, reusing, recycling, and remanufacturing, offers a pathway to a more sustainable future, one where economic development is no longer limited by the availability of raw materials but is instead driven by innovation, resource efficiency, and environmental stewardship (Desing et al., 2020; Perey et al., 2018).

3 Advancing Waste Reduction and Resource Efficiency Through CE

In the context of the CE, businesses are increasingly turning to innovative business models that prioritize waste reduction and resource optimization as fundamental pillars of sustainability (Perey et al., 2018; Rathore & Sarmah, 2020). These models extend the product lifecycle, create value from waste, and enable the continuous flow of materials within the economy (Romero-Hernández & Romero, 2018; Serrano-Arévalo et al., 2024). Unlike traditional linear business models, which assume a "take-make-dispose" approach, circular business models seek to close the loop, ensuring that resources are used efficiently, and waste is minimized at every stage of production and consumption (Chowdhury & Asiabanpour, 2024; Yang et al., 2024). This section discusses several circular business models that play key roles in advancing resource efficiency and waste reduction.

One prominent model is Product-as-a-Service (PaaS), which shifts from ownership-based consumption to service-based consumption (Tomić & Schneider, 2020; Prieto-Sandoval et al., 2019). In this model, businesses lease products to customers rather than selling them outright. The company retains ownership of the product throughout its lifecycle, which allows it to take responsibility for its maintenance, repair, and eventual return. This model encourages manufacturers to design durable and repairable products since they are responsible for maintaining their value over time. By focusing on longevity and reuse, the PaaS model not only reduces the frequency of product disposal but also facilitates the return of materials to the manufacturer for remanufacturing or recycling, thus minimizing waste and optimizing resource utilization (Gómez-Prado et al., 2022; Yang et al., 2024). Furthermore, this model reduces the demand for new resources as the same products and materials are reused multiple times (Hidalgo et al., 2019; Chioatto & Sospiro, 2023).

Another circular business model that fosters waste reduction and resource optimization is Extended Producer Responsibility (EPR) (Serrano-Arévalo et al., 2024; Ishaq et al., 2024). EPR shifts the responsibility for waste management from governments or consumers to producers. Under this model, manufacturers are held accountable for the entire lifecycle of their products, including take-back, recycling, or proper disposal once the products reach the end of their useful life. EPR incentivizes businesses to design products with easier disassembly, recyclability, and reusability in mind, ensuring that products can be efficiently returned to the supply chain (Tanveer et al., 2024; Prieto-Sandoval et al., 2019). This model reduces waste by encouraging producers to take responsibility for their products after they are sold and used, thus enhancing recycling rates and promoting the recovery of valuable materials (Perey et al., 2018; Baratsas et al., 2021). It also creates a market for second-hand materials, supporting circular material flows within industries (Tomić & Schneider, 2020).

Reverse logistics is another critical business model that supports the circular economy by enabling the efficient return of products and materials (Chowdhury & Asiabanpour, 2024; Hidalgo et al., 2019). In reverse logistics, companies set up systems to collect used products, components, and materials from customers for refurbishment, recycling, or reuse. Unlike traditional logistics, which focuses on the movement of goods from producer to consumer, reverse logistics is concerned with the movement of goods from consumers back to producers or recyclers. This model not only facilitates the recovery of valuable materials but also reduces the need for raw materials by reintroducing used products and components into the supply chain (Romero-Hernández & Romero, 2018; Rathore & Sarmah, 2020). It supports both waste reduction and resource optimization by ensuring that valuable resources are continuously circulated, reducing the need for new materials and lowering overall production costs (Tanveer et al., 2024; Prieto-Sandoval et al., 2019).

In addition to these models, there are other emerging circular business models that are making significant strides in resource efficiency and waste reduction. For example, the sharing economy or collaborative consumption model allows consumers to share products or services, which reduces the need for individual ownership and extends the product lifecycle (Najar et al., 2024; Rathore & Sarmah, 2020). Similarly, the remanufacturing model focuses on restoring used products to a like-new condition, reducing the need for new materials and minimizing waste (Hidalgo et al., 2019; Serrano-Arévalo et al., 2024). By bringing previously used products back into circulation, remanufacturing contributes to resource optimization and helps extend the life of products that might otherwise have been discarded (Perey et al., 2018; Rathore & Sarmah, 2020).

The integration of these models into business practices represents a shift toward a more sustainable and resource-efficient economic system. Each of these models has a distinct impact on waste reduction and resource optimization, contributing to the broader goals of the CE by promoting the reuse, recycling, and regeneration of materials (Tanveer et al., 2024; Prieto-Sandoval et al., 2019). These business models demonstrate the potential of the CE to address pressing environmental challenges by transforming the way businesses approach resource use and waste management (Hidalgo et al., 2019). By embracing these models, companies can contribute to a more sustainable and resource-efficient economy, where waste is minimized, resources are optimized, and the lifecycle of materials is extended. Moving forward, businesses will need to integrate these models into their strategies to capitalize on the economic, environmental, and social benefits that come from reducing waste and optimizing resource utilization (Romero-Hernández & Romero, 2018;

Table 1 Waste reduction and resource optimization aspects of circular business models

Business models	Impacts on waste reduction	References	Impact on resource optimization	References
Product-as-a-service (PaaS)	Reduces product disposal and waste generation by ensuring products are returned, repaired, or reused.	Rathore and Sarmah (2020); Najar et al. (2024); Chowdhury and Asiabanpour, (2024)	Maximizes resource use by retaining ownership and reusing materials over multiple cycles.	Gómez-Prado et al. (2022); Tanveer et al. (2024); Chioatto and Sospiro (2023).
Extended producer responsibility (EPR)	Encourages design for recycling and waste reduction by holding manufacturers accountable for post-consumer waste.	Serrano-Arévalo et al. (2024); Rathore and Sarmah (2020); Prieto-Sandoval et al. (2019)	Promotes material recovery, reducing the need for virgin materials by recycling and reusing components.	Perey et al. (2018); Baratsas et al. (2021); Tomić and Schneider, (2020)
Reverse logistics	Facilitates the recovery of products and materials from consumers, reducing waste and preventing landfills.	Chowdhury and Asiabanpour (2024); Hidalgo et al. (2019); Prieto-Sandoval et al. (2019)	Enables the recovery of valuable materials, ensuring that products are reused, remanufactured, or recycled into new products.	Romero-Hernández and Romero (2018); Rathore and Sarmah (2020); Tanveer et al. (2024); Delchet-Cochet, 2020
Sharing economy	Reduces waste by sharing products instead of ownership, decreasing the frequency of product disposal.	Keijer et al. (2019); Najar et al. (2024); Rathore and Sarmah (2020)	Optimizes resource utilization by extending the lifecycle of products through shared use.	Perey et al. (2018); Rathore and Sarmah, (2020); Chioatto and Sospiro (2023); Yang et al. (2024)
Remanufacturing	Reduces waste by refurbishing and restoring products to a like-new condition, avoiding the need for new raw materials.	Tanveer et al. (2024); Prieto-Sandoval et al. (2019); Romero-Hernández and Romero (2018)	Enhances resource efficiency by reducing the need for virgin materials and extending the life of existing resources.	Hidalgo et al. (2019); Romero-Hernández and Romero (2018); Prieto-Sandoval et al. (2019)

Source: Authors

Serrano-Arévalo et al., 2024; Yang et al., 2024). Table 1 highlights the waste reduction and resource optimization aspects of circular business models.

4 Theoretical Framework for Waste Reduction and Resource Optimization (WRRO)-Oriented Circular Business Model

In the context of organizational transformation, integrating waste reduction and resource optimization into business models is critical for successfully adopting the CE. This process entails redesigning how a company operates, interacts with customers, and manages its supply chain to focus on reducing waste and maximizing the efficient use of resources throughout the lifecycle of products and services (Tanveer et al., 2024; Romero-Hernández & Romero, 2018). This theoretical framework delves into the key criteria for designing and implementing circular business models that prioritize waste reduction and resource optimization.

4.1 Business Perspective for CE

The first step in integrating circularity into an organization is to reframe its business perspective to align with CE principles (Tanveer et al., 2024; Serrano-Arévalo et al., 2024). At the heart of this transformation is a shift in how value is created and delivered. Traditional business models focus primarily on profit generation through linear processes of resource extraction, production, consumption, and disposal (Perey et al., 2018; Rathore & Sarmah, 2020). In contrast, circular business models view value creation as a process that extends beyond simply selling products to include the management of their entire lifecycle, ensuring that products, materials, and resources are kept in circulation for as long as possible (Keijer et al., 2019; Chioatto & Sospiro, 2023). The value proposition in a circular business model should not only emphasize product quality and performance but also the sustainable use of resources, waste reduction, and the reuse of materials. Furthermore, customer relationships must evolve to encourage sustainable consumption patterns. CE models prioritize the idea of product-as-a-service, in which consumers access products rather than own them outright. This model reduces the frequency of product disposal and incentivizes companies to design durable, long-lasting, and repairable products (Tomić & Schneider, 2020; Tanveer et al., 2024). Instead of simply being a transactional relationship, the company–customer dynamic in a CE model shifts toward collaboration, with customers contributing to the sustainability efforts by returning products for remanufacturing or recycling. Resource flows within the organization must also be realigned to ensure that materials and products are continuously cycled through the production process, rather than being discarded after a single use (Romero-Hernández & Romero, 2018; Serrano-Arévalo et al., 2024). This rethinking of resource flows requires companies to rethink their entire value chain, including upstream processes (such as sourcing and production) and downstream processes (such as end-of-life management and waste recovery) (Tanveer et al., 2024; Perey et al., 2018). By integrating circularity into every step of the value chain, businesses can optimize resource use, minimize waste, and lower environmental impact, while also creating new opportunities for value creation.

4.2 Waste Reduction Strategies

The foundation of a waste reduction strategy in a circular business model is to design out waste at the very beginning of the product lifecycle (Serrano-Arévalo et al., 2024; Rathore & Sarmah, 2020). Waste reduction should not merely be an afterthought but should be embedded in the design, production, and consumption processes (Tanveer et al., 2024; Prieto-Sandoval et al., 2019). One of the core strategies is design for disassembly, which ensures that products can be easily taken apart to recover valuable materials, extend the life of individual components, and facilitate recycling or reuse. Lean manufacturing practices also play a crucial role in waste reduction. Lean principles focus on minimizing waste throughout production by optimizing processes, eliminating inefficiencies, and improving product quality (Chowdhury & Asiabanpour, 2024; Rathore & Sarmah, 2020). These strategies can significantly reduce material wastage, and the environmental footprint associated with overproduction, defects, and excess inventory (Keijer et al., 2019; Hidalgo et al., 2019). Additionally, adopting supply chain transparency allows organizations to track material flows and identify opportunities for waste reduction at every stage of the product lifecycle, from sourcing raw materials to end-of-life disposal. Another key strategy for waste reduction is resource recovery. This involves collecting and processing materials from used products, waste, or by-products, which can then be reintroduced into the production cycle (Romero-Hernández & Romero, 2018; Rathore & Sarmah, 2020), which not only reduces the amount of waste sent to landfills but also minimizes the demand for virgin materials, thereby conserving natural resources and reducing the overall environmental impact.

4.3 Waste-to-Value Strategies

A critical component of the CE is the ability to transform waste into value, converting discarded materials into useful resources (Perey et al., 2018; Rathore & Sarmah, 2020). Companies can adopt waste-to-value strategies, such as chemical recycling, where complex materials like mixed plastics are converted into new raw materials through chemical processes (Hidalgo et al., 2019; Tanveer et al., 2024), or waste-to-energy technologies, where non-recyclable waste is used to generate energy, either through combustion or conversion into biogas (Keijer et al., 2019; Romero-Hernández & Romero, 2018). These strategies reduce the environmental impact of waste by repurposing it into valuable materials or energy, thus preventing landfill accumulation and reducing the need for virgin resources (Najar et al., 2024; Serrano-Arévalo et al., 2024). Another innovative waste-to-value approach is upcycling, where waste materials are creatively repurposed into new products of higher value (Rathore & Sarmah, 2020; Prieto-Sandoval et al., 2019). By finding new uses for what would otherwise be discarded, businesses not only contribute to reducing waste but also create new revenue streams. These approaches also play a role in improving resource efficiency, as they help to conserve raw materials and reduce the carbon footprint of production (Tanveer et al., 2024; Hidalgo et al., 2019). As businesses develop waste-to-value solutions, it becomes important to integrate them into the broader CE ecosystem, where waste from one industry or sector becomes a valuable input for another (Keijer et al., 2019; Serrano-Arévalo et al., 2024). This fosters inter-industry collaboration and creates a network of businesses working together to optimize resource flows across multiple sectors, further reducing waste and maximizing the value derived from all materials.

4.4 Collaborative and Circular Supply Chains

In a circular business model, the supply chain is no longer a linear process where materials are sourced, used, and discarded. Instead, it becomes a collaborative network where stakeholders across the entire value chain work together to ensure that materials are recovered, reused, and recycled (Ngo et al., 2024; Perey et al., 2018). Collaboration across supply chains is critical for achieving resource optimization and waste reduction, as it ensures that waste is minimized, materials are continuously circulated, and resources are utilized more efficiently. For example, reverse logistics plays an essential role in circular supply chains by facilitating the collection and return of products from consumers for reuse, remanufacturing, or recycling (Chowdhury & Asiabanpour, 2024; Romero-Hernández & Romero, 2018; Ishaq et al., 2024). Suppliers, manufacturers, distributors, and recyclers need to work together to streamline these processes, reducing inefficiencies and ensuring that materials are properly recovered and reused (Najar et al., 2024; Hidalgo et al., 2019). Collaboration also extends to partnerships with third-party recyclers, technology providers, and waste management experts, all of whom are integral to creating a closed-loop supply chain (Perey et al., 2018; Rathore & Sarmah, 2020). These partnerships are vital for businesses looking to scale their circular practices, as they help ensure that the recovery, processing, and reuse of materials are carried out efficiently and effectively (Keijer et al., 2019; Chioatto & Sospiro, 2023). By fostering strong relationships across the value chain, companies can ensure that every link in the supply chain contributes to minimizing waste, improving resource efficiency, and creating value from waste. Figure 1 outlines the conceptualized WRRO circular model.

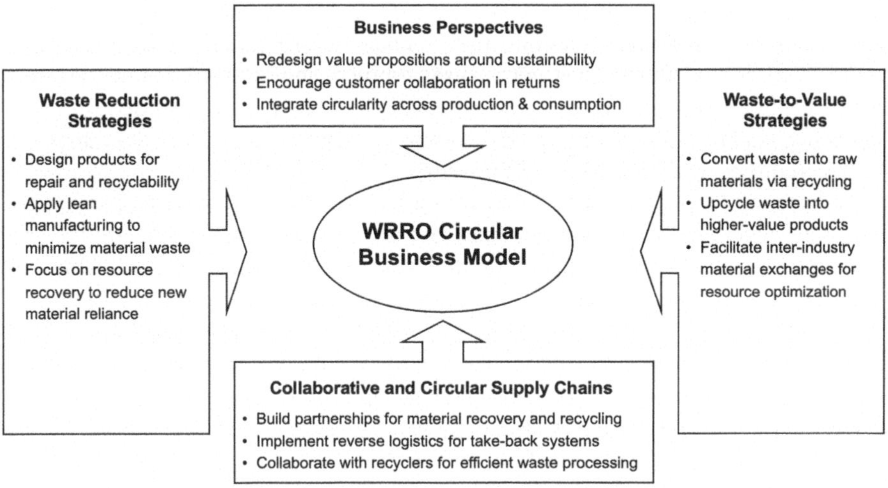

Fig. 1 The WRRO circular business model. *Source: Authors*

5 Case Study

In this case study, we analyze how X Corporation, a global manufacturing organization, has adopted CE principles to optimize resource utilization and reduce waste. X Corporation's approach is based on its EcoCycle Double Loop Framework, a comprehensive system introduced in 2004 to guide the company's transition toward a circular business model. The EcoCycle Framework integrates all stages of the product lifecycle—from material extraction to product disposal—focusing on sustainable practices across both upstream and downstream operations. Key elements of X Corporation's CE strategy include long-term product use, remanufacturing, material recycling, and the use of waste-to-value technologies, all aimed at minimizing waste and maximizing resource efficiency.

5.1 Case Background

In the face of growing environmental challenges, X Corporation has embarked on a transformative journey toward a sustainable future, driven by the principles of the CE. Since the mid-1990s, X Corporation has been a pioneer in reimagining how businesses can integrate sustainability into their operations, not just by minimizing environmental impact, but by creating a system where resources are used efficiently, waste is minimized, and products are kept in use for as long as possible. The company's commitment to a circular model is anchored in its EcoCycle Framework, a comprehensive approach to sustainability that emphasizes product longevity, material recovery, and continuous resource optimization. X Corporation's CE model is built around two interconnected loops designed to create a closed system of resource flow. The first loop focuses on collection, remanufacturing, and reuse, where used products are collected from customers, disassembled, and remanufactured to restore them to a like-new condition, ready for resale. This loop aims to extend product lifecycles, ensuring that materials and components are reused rather than disposed of, thus reducing the need for virgin resources.

The second loop is centered on recycling and material recovery, where components and materials that cannot be reused are processed and sent back into the production cycle. This loop facilitates the transformation of waste into valuable raw materials, ensuring that discarded products and materials are not wasted, but rather repurposed into new products or energy. Together, these two loops embody X Corporation's vision of a regenerative business model, where waste is minimized, and resource consumption is continuously optimized. Through its EcoCycle Framework, X Corporation not only demonstrates a commitment to sustainability but also creates a model for others to follow in integrating circular principles into their business strategies. The company's approach emphasizes collaboration across its supply chain, from material suppliers to recyclers, and fosters a culture of innovation where waste is viewed as a resource, and products are designed for reuse, recycling, and remanufacturing. This comprehensive CE framework has positioned X Corporation as a leader in sustainable business practices, driving both environmental and economic value. Figure 2 outlines the EcoCycle Framework at X Corporation.

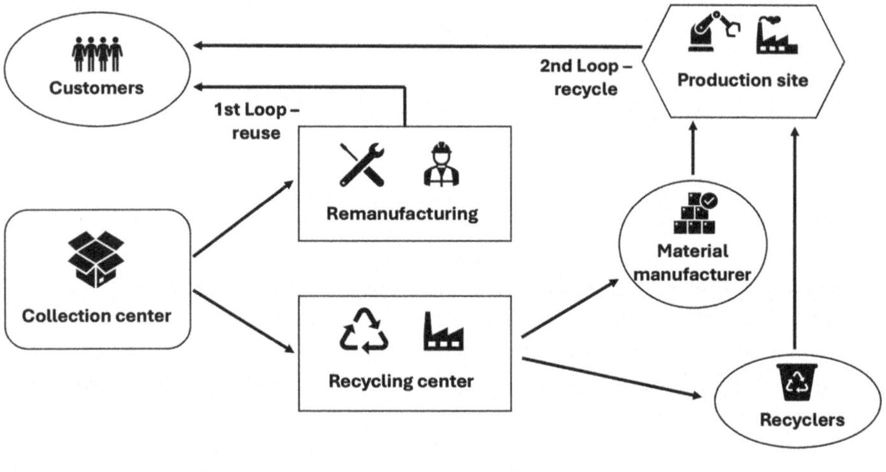

Fig. 2 EcoCycle framework at X Corporation. *Source: authors*

5.2 *Case Analysis*

This case study applies the theoretical framework for WRRO circular business models, focusing on four main criteria: Business Perspective for CE, Waste Reduction Strategies, Waste-to-Value Strategies, and Collaborative and Circular Supply Chains.

5.2.1 Business Perspective

X Corporation's adoption of the CE is framed around reframing value propositions, customer relationships, and resource flows. The company views sustainability as a core business strategy, not just an environmental objective. Its value proposition focuses on delivering high-quality products with long life cycles, which are returned, refurbished, and reused through a collection and remanufacturing system. This approach aligns with customer demand for environmentally responsible products and emphasizes the company's commitment to reducing environmental impact throughout the product lifecycle.

X Corporation has shifted its customer relationships by integrating product-as-a-service models, where customers lease products instead of purchasing them outright. This allows the company to maintain ownership, control the product lifecycle, and ensure products are returned for remanufacturing or recycling. This model reduces the need for new materials and promotes the reuse of existing resources, ensuring products remain within the company's resource loop. Additionally, the company has realigned resource flows within its operations to emphasize circularity, focusing on collection, recycling, and remanufacturing materials that would otherwise be discarded. This realignment reduces the organization's reliance on virgin resources, lowers production costs, and minimizes waste throughout its operations.

5.2.2 Waste Reduction Strategies

Waste reduction is a cornerstone of X Corporation's circular business model. The company implements several key waste reduction strategies, starting with design for longevity. Products are designed for durability, repairability, and recyclability, ensuring they can be reused and maintained for long periods. This design approach reduces the need for frequent product replacements and minimizes the disposal of outdated products. Additionally, X Corporation applies lean manufacturing techniques to reduce waste during production, including optimizing production processes, eliminating inefficiencies, and ensuring high-quality output with minimal defects.

Another essential waste reduction initiative is resource recovery. X Corporation has established robust systems for product take-back and reuse, collecting over 300,000 end-of-life products annually from customers around the world. The company disassembles these products to recover valuable components and materials, significantly reducing the volume of waste sent to landfills. This resource recovery process aligns with the company's zero waste goal, ensuring that materials are cycled back into the production process rather than being discarded.

5.2.3 Waste-to-Value Strategies

X Corporation has incorporated waste-to-value strategies into its operations, enabling the company to convert waste into valuable resources. One such strategy is chemical recycling, where waste plastics are converted into new raw materials for production. This process reduces reliance on virgin plastics and minimizes environmental impact by recycling materials that are typically difficult to process. Additionally, X Corporation uses waste-to-energy technologies to convert non-recyclable waste into thermal energy, reducing energy consumption and cutting costs.

The company's remanufacturing efforts further illustrate its waste-to-value strategies. X Corporation's remanufacturing centers disassemble used products, recover valuable parts, and restore them to like-new condition for resale. This process not only reduces the need for new raw materials but also ensures that products are reused to their fullest extent, preventing valuable resources from being discarded. By incorporating these waste-to-value strategies, X Corporation can derive value from waste, reduce its environmental impact, and improve resource efficiency.

5.2.4 Collaborative and Circular Supply Chains

A critical component of X Corporation's CE model is the establishment of collaborative and circular supply chains. The company recognizes that achieving true circularity requires the involvement of multiple stakeholders across the supply chain, including suppliers, manufacturers, recyclers, and customers. X Corporation collaborates closely with its suppliers to source materials that are low in carbon emissions and have minimal environmental impact. It also partners with recycling companies to ensure the efficient processing of used products and materials for reuse. By fostering these partnerships, X Corporation ensures that materials are recovered and reused in a way that benefits the entire supply chain.

X Corporation's reverse logistics system plays an essential role in its circular supply chain by enabling the efficient return of products and materials. The company has implemented a global take-back program, allowing customers to return used products for reuse, recycling, or remanufacturing. This program significantly reduces waste by ensuring that products are not discarded but instead returned to the production cycle for further use. X Corporation also works with waste disposal partners certified to handle waste in an environmentally responsible manner. By building a strong network of partners across its supply chain, X Corporation ensures that materials are efficiently recovered, waste is minimized, and sustainability is prioritized throughout its operations.

X Corporation's approach to the CE, as exemplified through its EcoCycle framework, demonstrates how businesses can effectively integrate waste reduction and resource optimization into their operations and supply chains. Through innovative business models such as product-as-a-service, extended producer responsibility, and reverse logistics, X Corporation has successfully minimized waste, optimized resource use, and created value from discarded materials. The company's commitment to collaboration across its supply chain further strengthens its circular business model, ensuring that sustainability is integrated into every step of the production, consumption, and disposal process. This case study illustrates the practical application of CE principles in reducing environmental impact while improving business efficiency and creating new economic value.

6 Challenges in Implementing Waste Reduction and Resource Optimization (WRRO)-Oriented Circular Business Models

Implementing a WRRO-Oriented Circular Business Model involves significant shifts in how businesses operate, engage with customers, and interact with their supply chains. While the potential benefits are substantial, such as reduced environmental impact and increased operational efficiency, there are numerous challenges in realizing these goals. These challenges span several dimensions, including business strategy, product design, operational processes, and collaboration across the value chain. Below, we explore the specific difficulties associated with each dimension of the WRRO model.

6.1 Business Perspective for CE

Shifting business models from linear to circular requires a reframing of value propositions and resource flows that are not always easy to implement (Delchet-Cochet, 2020; Perey et al., 2018). One key difficulty is convincing organizations to move away from traditional ownership-based models to service based models, such as product-as-a-service. In such models, consumers lease products rather than own them, which requires the company to maintain control over the product's lifecycle. This shift can be met with resistance from both customers, who may not understand the value of leasing over ownership, and businesses, which must adjust their pricing models, profit structures, and operations (Romero-Hernández & Romero, 2018; Prieto-Sandoval et al., 2019; Ishaq et al., 2024). Furthermore, integrating sustainability into value propositions often requires organizations to rethink their relationships with customers. Businesses must work collaboratively with consumers, encouraging them to return used products, repair, and reuse items (Rathore &

Sarmah, 2020; Hidalgo et al., 2019). This can be a difficult transition, especially in consumer markets that have long relied on linear consumption patterns, where products are discarded once they lose value or functionality. This shift also requires companies to clearly communicate the long-term value of circular models, including environmental benefits and potential cost savings, to both consumers and other stakeholders.

6.2 *Waste Reduction Strategies*

One of the core elements of the WRRO-oriented model is waste reduction, achieved through strategies such as design for longevity, material recovery, and recycling (Desing et al., 2020; Perey et al., 2018). The challenges here stem from the complexity and cost associated with designing products for longevity and recyclability. Manufacturers must rethink the way products are designed, shifting from disposable products to those that can be easily repaired, disassembled, and recycled. This requires investment in product redesign, materials innovation, and the development of modular components—all of which can increase product costs and complicate manufacturing processes (Tomić & Schneider, 2020; Prieto-Sandoval et al., 2019). Moreover, adopting lean manufacturing techniques to minimize material waste during production is essential, but this too can be challenging. Companies must optimize their production processes, often requiring new technologies, employee training, and more efficient supply chain management (Perey et al., 2018; Serrano-Arévalo et al., 2024; Sheikh et al., 2024). Waste minimization strategies, such as reducing scrap, avoiding overproduction, and reusing waste materials, require businesses to adopt new metrics and practices (Romero-Hernández & Romero, 2018; Rathore & Sarmah, 2020). This process can also be costly in the short term, as it often requires upfront investment in new technology or a shift in existing operations to streamline waste reduction efforts.

6.3 *Waste-to-Value Strategies*

Converting waste into value, through strategies such as chemical recycling and waste-to-energy technologies, is a key challenge for companies seeking to optimize resource use (Keijer et al., 2019; Rathore & Sarmah, 2020). Chemical recycling, for example, involves transforming waste materials like plastics into new raw materials for use in production. However, this process is still technologically complex and not universally available (Desing et al., 2020; Tomić & Schneider, 2020). The challenges associated with scaling up chemical recycling include high operational costs, technological limitations, and a lack of infrastructure to process materials efficiently. Similarly, waste-to-energy technologies face challenges related to the availability of suitable waste materials, the cost-effectiveness of the process, and concerns over the environmental impact of converting waste into energy (Romero-Hernández & Romero, 2018; Serrano-Arévalo et al., 2024). In some industries, the market for waste-to-energy is still underdeveloped, and companies may not have access to affordable or sustainable methods for converting waste to energy (Chioatto & Sospiro, 2023; Prieto-Sandoval et al., 2019). Furthermore, waste-to-energy solutions may not always align with broader sustainability goals if the energy recovery processes themselves result in emissions or other environmental concerns (Perey et al., 2018; Rathore & Sarmah, 2020). The key challenge in waste-to-value strategies is also the difficulty in establishing consistent, high-quality waste streams that can be used in the manufacturing process (Serrano-Arévalo et al., 2024; Rathore & Sarmah, 2020). Waste materials can vary significantly in quality, and finding reliable ways to repurpose them into high-value products or resources is not always feasible without the right technology, infrastructure, or market acceptance.

6.4 Collaborative and Circular Supply Chains

The implementation of WRRO-oriented business models also requires extensive collaboration across supply chains. A circular supply chain is inherently more complex than traditional linear models because it requires stakeholders from various industries, including suppliers, manufacturers, recyclers, logistics providers, and consumers, to work together (Romero-Hernández & Romero, 2018; Chioatto & Sospiro, 2023). Coordinating these partners to ensure that materials are collected, reused, and recycled efficiently is one of the greatest challenges faced by businesses seeking to adopt circular practices (Serrano-Arévalo et al., 2024; Hidalgo et al., 2019). Reverse logistics is one of the most critical components of a circular supply chain, yet it can be extremely difficult to manage. Collecting used products and ensuring they are properly returned, sorted, and sent to the right destinations (for reuse, remanufacturing, or recycling) involves significant logistical complexity (Keijer et al., 2019; Rathore & Sarmah, 2020; Ishaq et al., 2024). Developing efficient systems for product returns, managing product take-back programs, and creating incentives for consumers to return products for reuse can be costly and operationally difficult (Chowdhury & Asiabanpour, 2024; Prieto-Sandoval et al., 2019). In many cases, businesses may not have access to the necessary infrastructure to support reverse logistics at scale, especially in global markets where logistical coordination and transportation systems vary widely (Perey et al., 2018; Najar et al., 2024). Additionally, creating a circular supply chain often requires forging new partnerships with recyclers, waste management companies, and other external stakeholders (Serrano-Arévalo et al., 2024; Luthra et al., 2022; Ngo et al., 2024). Aligning the interests of these partners, ensuring they meet the same sustainability standards, and ensuring transparency in the flow of materials can be difficult. Companies may face resistance from suppliers or partners who are not fully committed to circular principles or who lack the capabilities to meet CE standards (Romero-Hernández & Romero, 2018; Hidalgo et al., 2019). Furthermore, the complex nature of circular supply chains makes them difficult to scale, especially for large corporations with diverse product portfolios and international operations. Table 2 summarizes the findings and challenges for generalization.

Table 2 Research summary

Themes	Theoretical criteria	X Corporation's findings	Challenge for generalization
Business perspective	Redesign value propositions around sustainability. Encourage customer collaboration in recycling and returns. Integrate circular practices across production and consumption.	• Sustainability is central to its value proposition, focusing on long-term product use and environmental impact reduction. • Customer collaboration is encouraged for product returns and recycling, ensuring products stay within the resource loop. • The company integrates circular resource flows, focusing on reuse and remanufacturing.	• The transformation can face resistance from consumers and businesses. • Reframing value propositions around sustainability requires clear communication and market adaptation. • Customer collaboration for product returns and recycling is challenging due to existing consumption habits.
Waste reduction strategies	Design products for repair and recyclability. Apply lean manufacturing to minimize material waste. Focus on resource recovery to reduce new material reliance.	• X corporation emphasizes design for longevity, creating durable, repairable, and recyclable products. • Lean manufacturing is applied to optimize production processes, minimizing material waste. • Product take-back systems in place, reducing waste and encouraging recycling and reuse.	• Design for longevity and recyclability increases product complexity and production costs. • Lean manufacturing requires process optimization and resource efficiency, which can involve upfront investment. • Resource recovery and minimizing waste during production require new infrastructure and technologies, adding operational costs.

(continued)

Table 2 (continued)

Themes	Theoretical criteria	X Corporation's findings	Challenge for generalization
Waste-to-value strategies	Convert waste into raw materials via recycling. Use waste-to-materials technologies to generate energy from waste. Facilitate inter-industry material exchanges for resource optimization.	• Chemical recycling and waste-to-energy technologies are used to recycle waste plastics and convert non-recyclable materials into thermal energy. • X corporation implements remanufacturing, recovering parts and materials from used products and restoring them to like-new condition. • Inter-industry material exchange is facilitated, ensuring that waste from one sector serves as a resource for another.	• Chemical recycling is costly and technologically complex, with limited infrastructure. • Waste-to-energy technologies can face concerns over cost-effectiveness, environmental impact, and efficiency. • Establishing consistent waste streams for reuse can be difficult due to variability in material quality and availability.
Collaborative and circular supply chains	Build partnerships for material recovery and recycling. Implement reverse logistics for product take-back systems. Collaborate with recyclers for efficient waste processing.	• X Corporation's reverse logistics system ensures efficient product take-back from customers for reuse, remanufacturing, or recycling. • Collaborates with recyclers and suppliers to ensure the recovery and recycling of materials. • Works with waste disposal partners to manage waste responsibly and certified recycling companies for processing.	• Reverse logistics presents logistical complexities and operational costs for product take-back and sorting. • Supply chain coordination requires extensive partnerships, alignment of interests, and transparency across multiple stakeholders. • Scalability of circular supply chains is challenging, especially in global markets with diverse infrastructure and regulations.

Source: Authors

7 Concluding Remark

7.1 *Research Contributions*

This research enhances the growing body of literature on CE by offering a refined understanding of how businesses can transition to circular business models and optimize resource utilization. While existing studies have examined the benefits of CE principles, particularly in waste reduction and environmental sustainability (Delchet-Cochet, 2020; Rathore & Sarmah, 2020), this work deepens the theoretical framework by positioning resource efficiency as a central driver in business strategy, extending beyond its typical role as a secondary benefit (Chioatto & Sospiro, 2023; Hidalgo et al., 2019). Existing literature on CE models often focuses on material reuse and waste reduction (Keijer et al., 2019; Tomić & Schneider, 2020). However, this study shifts the emphasis to resource efficiency as a critical component of the CE. It underscores how optimizing resource use throughout the product lifecycle—encompassing design, production, consumption, and end-of-life management—offers a more comprehensive approach to circularity (Serrano-Arévalo et al., 2024; Rathore & Sarmah, 2020). This contribution expands current understandings of sustainable business practices by incorporating systemic resource efficiency at every stage of production and consumption, making it a core element rather than an ancillary consideration.

Moreover, this study advances the theoretical understanding of circular business models by introducing a dual-loop framework for resource flow. Traditional CE models often focus on a single resource flow, such as closed-loop recycling or product life extension (Perey et al., 2018; Prieto-Sandoval et al., 2019). In contrast, the dual-loop framework integrates both closed-loop (internal) and open-loop (external) systems, allowing businesses to manage and optimize resource flows across a broader spectrum (Chowdhury & Asiabanpour, 2024; Serrano-Arévalo et al., 2024). This holistic approach facilitates a more dynamic and scalable model, promoting environmental and economic sustainability (Keijer et al., 2019; Najar et al., 2024). It provides a more robust understanding of how circular practices can be operationalized across industries, allowing companies to achieve greater resource optimization and minimize waste.

Furthermore, this research challenges the prevailing notion that CE adoption is driven primarily by regulatory pressures or consumer demand (Tomić & Schneider, 2020; Rathore & Sarmah, 2020). Instead, it positions resource efficiency as a key business driver, particularly in industries where raw material costs fluctuate or supply chains are under strain (Serrano-Arévalo et al., 2024; Romero-Hernández & Romero, 2018; Ngo et al., 2024). By framing resource efficiency as both an environmental necessity and an economic imperative, this study offers a strategic perspective that integrates both cost-effectiveness and sustainability into the business case for CE models (Delchet-Cochet, 2020; Perey et al., 2018). This shift enriches the existing literature (such as Serrano-Arévalo et al., 2024; Rathore & Sarmah, 2020), which has traditionally emphasized the environmental or waste management aspects of circularity, by introducing the idea that resource efficiency is central to the business rationale for adopting circular practices.

On a practical level, this research provides actionable insights for businesses aiming to implement CE principles. The dual-loop framework offers not only a theoretical model for understanding resource flows but also strategic recommendations for businesses. By emphasizing partnerships, reverse logistics, and innovative design, the research outlines concrete pathways for businesses to improve resource efficiency and transition from linear to circular models without incurring significant costs or disrupting existing operations. These practical contributions directly inform the literature by providing a structured, actionable roadmap for businesses seeking to incorporate circular practices into their business models.

7.2 Future Research Suggestions

While this research contributes valuable insights into the application of CE principles for optimizing resource efficiency, several avenues for future research remain open.

One promising area for future research is the integration of digital technologies CE practices. This study primarily focuses on the theoretical models and strategic frameworks, but there is a growing need to investigate how technologies like Internet of Things (IoT), blockchain, AI, and big data can be leveraged to optimize resource flows and improve the efficiency of circular business models. Digital platforms could play a critical role in tracking product lifecycles, optimizing reverse logistics, and providing real-time data on material flows and waste reduction. Research could focus on how digital tools can enhance transparency in the circular supply chain, improve product traceability, and enable more efficient resource management.

While this study provides a general framework for CE implementation, there is a need for more industry-specific research to understand how CE principles can be tailored to sectors. For instance, industries such as electronics, fashion, and construction face unique challenges related to material recovery, product design, and supply chain logistics. Future research could explore how CE models can be customized for these industries, considering the specific technological, regulatory, and market conditions they operate under. Comparative studies across different sectors could provide valuable insights into the scalability and adaptability of CE strategies.

A key area for future research is the economic impact of transitioning from linear to circular business models. While this study emphasizes the strategic and resource efficiency benefits, empirical research is needed to assess the financial implications for businesses adopting circular practices. Future studies could investigate the costs and savings associated with circular transitions, including initial investment requirements, supply chain adjustments, and long-term cost savings from resource recovery, waste reduction, and reduced dependency on raw materials. A more detailed analysis of the return on investment (ROI) from circular practices would help businesses better understand the economic feasibility and benefits of adopting such models.

As CE models continue to evolve, it is essential to investigate the social implications of such transitions, particularly in terms of employment patterns, job creation, and labor market dynamics. Research could examine how shifts toward circular business models influence the demand for specific skills in industries such as remanufacturing, recycling, and product repair. This includes studying the potential for circular practices to generate green jobs and the role of education and training in preparing the workforce for the new demands of circular industries. Exploring the social equity implications, such as how CE practices can benefit or challenge different communities, is also a key area for future research.

Finally, longitudinal research is needed to assess the long-term outcomes of adopting CE practices. While this study provides a snapshot of the theoretical implications and strategic recommendations, it would be valuable to track the evolution of circular models over time, identifying the barriers, successes, and transformations businesses experience as they move toward circularity. Longitudinal studies can offer deeper insights into how businesses evolve in their journey toward circular practices and help inform best practices and roadmaps for other businesses looking to make the transition.

References

Baratsas, S. G., Pistikopoulos, E. N., & Avraamidou, S. (2021). A systems engineering framework for the optimization of food supply chains under circular economy considerations. *Science of the Total Environment, 794*, 148726.

Chioatto, E., & Sospiro, P. (2023). Transition from waste management to circular economy: The European Union roadmap. *Environment, Development and Sustainability, 25*(1), 249–276.

Chowdhury, H., & Asiabanpour, B. (2024). A circular economy integration approach into vertical farming with computer-based simulation model for resource optimization and waste reduction. *Journal of Cleaner Production, 470*, 143256.

Delchet-Cochet, K. (Ed.). (2020). *Circular economy: From waste reduction to value creation*. John Wiley & Sons.

Desing, H., Brunner, D., Takacs, F., Nahrath, S., Frankenberger, K., & Hischier, R. (2020). A circular economy within the planetary boundaries: Towards a resource-based, systemic approach. *Resources, Conservation and Recycling, 155*, 104673.

Gómez-Prado, R., Alvarez-Risco, A., Sánchez-Palomino, J., de las Mercedes Anderson-Seminario, M., & Del-Aguila-Arcentales, S. (2022). Circular economy for waste reduction and carbon footprint. In *Circular economy: Impact on carbon and water footprint* (pp. 139–159). Springer Singapore.

Hidalgo, D., Martín-Marroquín, J. M., & Corona, F. (2019). A multi-waste management concept as a basis towards a circular economy model. *Renewable and Sustainable Energy Reviews, 111*, 481–489.

Ishaq, S., Hoang, T. G., Tanveer, U., Hoang, T. H., & Truong, H. Q. (2024). Transformative capabilities of MedTech organizations in driving circularity in the healthcare industry: Insights from multiple cases. *Journal of Cleaner Production, 446*, 141370.

Keijer, T., Bakker, V., & Slootweg, J. C. (2019). Circular chemistry to enable a circular economy. *Nature Chemistry, 11*(3), 190–195.

Luthra, S., Mangla, S. K., Sarkis, J., & Tseng, M. L. (2022). Resources melioration and the circular economy: Sustainability potentials for mineral, mining and extraction sector in emerging economies. *Resources Policy, 77*, 102652.

Najar, I. N., Sharma, P., Das, R., Tamang, S., Mondal, K., Thakur, N., et al. (2024). From waste management to circular economy: Leveraging thermophiles for sustainable growth and global resource optimization. *Journal of Environmental Management, 360*, 121136.

Ngo, V. M., Quang, H. T., Hoang, T. G., & Binh, A. D. T. (2024). Sustainability-related supply chain risks and supply chain performances: The moderating effects of dynamic supply chain management practices. *Business Strategy and the Environment, 33*(2), 839–857.

Perey, R., Benn, S., Agarwal, R., & Edwards, M. (2018). The place of waste: Changing business value for the circular economy. *Business Strategy and the Environment, 27*(5), 631–642.

Prieto-Sandoval, V., Jaca, C., Santos, J., Baumgartner, R. J., & Ormazabal, M. (2019). Key strategies, resources, and capabilities for implementing circular economy in industrial small and medium enterprises. *Corporate Social Responsibility and Environmental Management, 26*(6), 1473–1484.

Rathore, P., & Sarmah, S. P. (2020). Economic, environmental and social optimization of solid waste management in the context of circular economy. *Computers & Industrial Engineering, 145*, 106510.

Romero-Hernández, O., & Romero, S. (2018). Maximizing the value of waste: From waste management to the circular economy. *Thunderbird International Business Review, 60*(5), 757–764.

Serrano-Arévalo, T. I., Díaz-Alvarado, F. A., Tovar-Facio, J., & Ponce-Ortega, J. M. (2024). The impact of circular economy indicators in the optimal planning of energy systems. *Sustainable Production and Consumption, 44*, 234–249.

Sheikh, A., Simske, S. J., & Chong, E. K. (2024). Evaluating artificial intelligence models for resource allocation in circular economy digital marketplace. *Sustainability, 16*(23), 10601.

Tanveer, U., Hoang, T. G., Truong, H. Q., Ishaq, S., & Gong, Y. (2024). The critical role of procurement in the emergence of circular business models: Insights from multiple cases of Vietnamese manufacturers. *Business Strategy and the Environment, 33*(8), 7689–7707. https://doi.org/10.1002/bse.3888

Tomić, T., & Schneider, D. R. (2020). Circular economy in waste management–socio-economic effect of changes in waste management system structure. *Journal of Environmental Management, 267*, 110564.

Whalen, K. A., & Whalen, C. J. (2018). The circular economy and institutional economics: Compatibility and complementarity. *Journal of Economic Issues, 52*(3), 605–614.

Yang, C. H., Kuo, L., Liu, Y. Y., & Pan, K. L. (2024). Incorporating resource optimization for sustainable airline service innovation business decision model: Toward circular economy policy achievement. *Annals of Operations Research*, 1–27.

Impact of Financial Development on Economic Growth in Developing Countries

Mai Nguyen Ngoc, Anna Maria Majo, and An Duong Thi Binh

Abstract This study provides significant new evidence on the relationship between financial development and economic growth in developing countries, using the generalized least squares (GLS) method for analysis. The research focuses on a sample of 30 developing countries worldwide, covering the period from 2010 to 2023. The study's objective is to examine how financial development influences economic growth in developing economies, which face unique challenges compared to developed countries. The empirical results demonstrate that the development of the banking sector and private sector credit positively affect economic growth. In contrast, liquid liabilities have a negative impact, and stock market development shows no significant effect. Second, government expenditures and inflation rates negatively affect economic growth in these countries. Third, trade openness plays an important role in fostering economic growth in the sample of developing countries. These findings offer valuable policy implications for governments in developing countries to enhance economic growth. Increasing bank branches and private sector credit can improve access to capital and investment. However, fluctuations in liquidity debt negatively correlate with growth, emphasizing the need for cautious monetary policies, improved debt management, and economic diversification. Developing countries should also focus on enhancing transparency, strengthening regulatory frameworks, and adopting advanced technologies such as blockchain and artificial intelligence to improve market efficiency and attract foreign investment.

M. N. Ngoc
Hanoi Foreign Trade University, Hanoi, Vietnam
e-mail: ngocmn@ftu.edu.vn

A. M. Majo
Marian College, Kuttikkanam, India

A. D. T. Binh (✉)
CIRTech Institute, HUTECH University, Ho Chi Minh City, Vietnam
e-mail: dtb.an@hutech.edu.vn

A. Sharma et al. (eds.), *The Synergistic Effect of Sustainable Business Practices on Corporate Performance*, https://doi.org/10.1007/978-981-96-3375-3_11

1 Introduction

Financial development encompasses a set of factors, policies, and institutions aimed at ensuring the deepening (including scale and liquidity), accessibility of capital and financial services (for individuals and businesses), as well as the efficiency (providing financial services at low cost, sustainable revenue) of financial intermediaries and markets (Čihák et al., 2012). Researchers have developed various indicators to measure financial development. For instance, domestic credit to the private sector and stock market capitalization are two commonly used indicators to represent financial depth. Meanwhile, the number of bank branches per 100,000 people and the proportion of market capitalization outside the top 10 largest companies in total stock market capitalization are used to measure the accessibility of the financial system.

Economic growth signifies the rise in producing goods and services, enhancing societal welfare. It is a crucial indicator of a nation's economic progress, usually measured by Gross Domestic Product (GDP). Economic growth demonstrates how economic activities increase community income over time by utilizing production factors to generate output. Economic growth factors include capital, human and natural resources, trade openness, government spending, and inflation.

Numerous theoretical and empirical studies have demonstrated the positive impact of financial development on countries' economic growth. When financial institutions and markets function effectively, they promote economic growth by reducing transaction costs, accelerating transactions, improving capital accumulation, and allocating investments to the most efficient sectors of the economy (Christopoulos & Tsionas, 2004; Estrada et al., 2010; Jedidia et al., 2014). Additionally, some studies have examined the role of financial sector development in relation to economic growth, specifically demonstrating the positive correlation between the development of the banking system and the economic growth of countries (Zhang et al., 2012; Pradhan et al., 2017).

Some studies show no correlation between financial market development and economic growth, while others indicate a negative impact. A complex financial system can lead to financial crises (Rajan, 2005), and an oversized system may cause resource misallocation. Recent studies suggest a nonlinear relationship, where the positive effect of financial development on growth can diminish or even reverse after a certain threshold (Demetriades & Rousseau, 2016).

Developing countries face significant financial challenges, mainly relying on banking systems for medium and long-term capital. This reliance creates risks for credit institutions, as their primary role is to provide short-term capital. Although corporate bond markets are vital in developed nations, many developing countries exhibit lax management and limited investor understanding of bonds. Furthermore, stock market growth is slow, failing to meet capital needs and exhibiting several negative traits. Therefore, creating a safe and sustainable financial system is essential for achieving economic development goals in these nations.

It is evident that the relationship between financial development and economic growth remains a complex topic with inconclusive findings. The nature of this relationship heavily depends on the context and characteristics of each country or group of countries being studied, leading to inconsistent empirical research results. Based on these theoretical and practical issues, this study examines the impact of financial development on the economic growth of 30 developing countries across four continents during the period from 2014 to 2023. Consequently, the study proposes policy implications to promote financial development and achieve economic growth objectives for developing countries.

2 Literature Review

2.1 Positive Impacts

Khan and Senhadji (2003) highlighted the significant role of financial development in economic growth through their study of 159 countries from 1960 to 1999, using 2SLS (Two-Stage Least Squares) and OLS (Ordinary Least Squares) estimation methods. Additionally, Ciftci et al. (2017), through AMG (Augmented Mean Group) & CCE (Common-Correlated Effects) analysis, concluded that credit from the financial market and equity from the stock market are two determinants of long-term per capita GDP in the 40 countries examined in their study.

Moreover, studies on the impact of financial development on economic growth emphasize each country's contextual and conditional factors. Rousseau and Wachtel (2002) found that financial development positively impacts countries with low inflation, while its effect is negligible in high-inflation environments. Jude (2010) and Huang and Lin (2009) indicated that the effect of financial development depends on factors such as trade openness, government consumption, inflation rates, and stages of economic development, with a stronger influence in low-income countries. Caporale et al. (2015) affirmed that the degree of influence varies according to a country's financial structure, while Ibrahim and Alagidede (2018) emphasized the necessity of sustainable financial management to support long-term growth, particularly in sub-Saharan Africa.

2.2 Negative Impacts

Using data from northern and southern Mediterranean countries, Ayadi et al. (2015) examined the impact of financial sector development on economic growth. The results of this study revealed a negative correlation between private sector credit and bank deposits with economic growth. Similarly, Arcand et al. (2015) identified the diminishing returns of financial development to economic growth. According to the authors, when private sector credit reaches a certain threshold, financial development's impact on output growth decreases and becomes negative. Demetriades and Rousseau (2016) demonstrated that the influence of financial development diminishes over time, and its long-term effect is negligible. Their study also highlighted banking systems' regulatory and supervisory role in moderating the relationship between financial development and growth.

2.3 Causal Relationship

A study by Grassa and Gazdar (2014) explored the impact of Islamic financial development and conventional financial development on economic growth in five Gulf Cooperation Council (GCC) countries: Bahrain, Kuwait, Qatar, Saudi Arabia, and the United Arab Emirates (UAE). The empirical findings indicated no correlation between conventional financial development and growth. However, the authors discovered that Islamic finance leads to regional growth. On the other hand, Yang's (2019) research on 49 countries from 1970 to 2016 found evidence of a Granger causal relationship between stock market development and economic growth across all three groups of economies. However, the strength of this relationship varied. Using a VAR model, the study also showed a reverse causal relationship between economic growth and capital market development in high-income economies, which was not observed in other economies.

2.4 Studies in Developing Countries

Research on the relationship between financial development and economic growth in developing countries has yielded notable findings. Christopoulos and Tsionas (2004) identified a long-term, unidirectional causal relationship between financial development and economic growth. Estrada et al. (2010) confirmed the critical role of bank credit, liquid debt, and stock markets in economic growth, particularly during the Asian financial crisis. Hassan et al. (2011) and Narayan and Narayan (2013) also found a positive relationship, where bank credit significantly contributes to economic growth, though the impact varies across regions. Jedidia et al. (2014) and Pradhan et al. (2017) emphasized the role of stock markets, the banking sector, and insurance in promoting long-term growth. In Vietnam and Southeast Asia, Ho (2018) and Trinh (2018) demonstrated a bidirectional causal relationship between financial development and economic growth, with varying effects depending on financial indicators. These findings underscore the essential role of financial development in driving economic growth in developing countries.

2.5 Research Gap

This study addresses key research gaps. First, examining the impact of recent financial development will provide practical conclusions relevant to current contexts. Existing studies mainly analyze the general relationship between financial development and economic growth, often neglecting specific channels of influence, such as bank credit, stock markets, central banks, and non-bank financial institutions. Developing countries feature diverse financial systems, including credit markets, banking, and stock markets. Understanding these different financial structures will support more effective financial policies in each country. Finally, this study's diverse range of developing countries allows for broader conclusions than previous research with limited geographic focus countries.

3 Methodology

The relationship between financial development and economic growth has been addressed in the endogenous growth theory of Romer (1986) and tested through the results of numerous empirical studies. To evaluate financial development in correlation with economic growth, researchers primarily use indicators of depth, accessibility, and efficiency of the financial system and financial markets as proposed by the IMF (2012). Additionally, macroeconomic variables are included in the models as moderators of the finance-growth relationship. Based on the research models proposed by Estrada et al. (2010) and Le and Nguyen (2021), this study proposes the following research model:

$LGDP_{it} = \beta_0 + \beta_1 LIQ_{it} + \beta_2 BAN_{it} + \beta_3 PRC_{it} + \beta_4 STO_{it} + \beta_5 GEX_{it} + \beta_6 TRA_{it} + \beta_7 INR_{it} + c_i + u_{it}$.

In this model, the dependent variable **economic growth (LGDP)** is the country's annual GDP per capita logarithmic value. Financial development is measured by four factors: the ratio of liquid liabilities to GDP (LIQ), banking system development (BAN), represented by the number of bank branches per 100,000 people, the ratio of private credit to GDP (PRC), and the market capitalization of the stock market as a percentage of GDP (STO). The control variables in the model include government expenditure as a percentage of GDP (GEX), trade openness (TRA), measured as the ratio of total trade volume to GDP, and the inflation rate (INR).

3.1 Explanation of Variables

Liquid Liabilities Liquid liabilities (M3) include total currency (M0), transferable deposits (M1), term deposits, savings, and other financial instruments (M2), as well as traveler's checks, commercial papers, and mutual fund shares. The ratio of liquid liabilities to GDP measures financial depth and is frequently used in studies on financial development, showing a positive impact on economic growth. Empirical studies by Mishra et al. (2009), Estrada et al. (2010), Le and Nguyen (2021), and Hui et al. (2021) have confirmed the positive relationship between the ratio of liquid liabilities to GDP and economic growth.

Banking System Development A sustainable and developed banking system provides economic benefits while minimizing negative social and environmental impacts. The number of bank branches per 100,000 people is often used to evaluate access to the financial system. Empirical studies also support the positive relationship between banking system development and economic growth in various countries (Estrada et al., 2010; Zhang et al., 2012; Demetriades & Rousseau, 2016; Pradhan et al., 2017).

Private Sector Credit Private sector credit includes loans to non-financial enterprises, households, and non-profit organizations. The ratio of private credit to GDP reflects the level of credit available to the private sector. A high ratio indicates growth in credit, greater access to capital, and higher investment levels, demonstrating the development of the financial system and its contribution to economic growth. Numerous studies on financial development and its correlation with economic growth have established a positive relationship between private sector credit and economic growth (Bayar, 2014; Jedidia et al., 2014; Ho, 2018; Le & Nguyen, 2021).

Stock Market Development The stock market is an effective capital mobilization channel for businesses, increasing investment resources, business expansion, job creation, and economic growth. Stock market development is commonly measured by the market capitalization-to-GDP ratio, which reflects the financial depth of the economy. Empirical studies by Jedidia et al. (2014), Pradhan et al. (2017), and Bist (2018) have demonstrated the positive impact of stock market development on economic growth.

Government Expenditure The ratio of government expenditure to GDP measures the final government consumption relative to GDP. This includes public consumption, the provision of goods and services, and public investment in infrastructure, education, and healthcare. Government expenditure promotes economic growth by providing public goods, supporting markets, creating positive externalities, and increasing labor productivity, as demonstrated in empirical studies (Jude, 2010; Arcand et al., 2015; Kimaro et al., 2017). However, government spending can also hinder economic growth due to the negative effects of taxation, borrowing, or inefficiencies (Trinh, 2018).

Trade Openness Trade openness, measured as the total trade volume (exports + imports) relative to GDP, fosters economic growth through three channels: (i) efficient resource allocation, (ii) technology transfer from developed to developing countries, and (iii) innovation in developed countries applied by developing nations, according to theories by Romer (1986) and Hye (2012). The positive impact of trade openness on economic growth has also been found in empirical studies by Hassan et al. (2011), Belloumi (2014), Demetriades and Rousseau (2016), and Le and Nguyen (2021). However, some studies have also highlighted the negative effects of trade openness on growth.

Inflation Measured by the annual consumer price index, inflation significantly impacts economic growth. High inflation increases costs, stagnates production, and hampers growth, while moderate inflation can foster economic development. Empirical research shows that the relationship between inflation and economic growth varies across countries. Studies by Christopoulos and Tsionas (2004) and Salahuddin and Gow (2016) found that inflation negatively affects growth. However, some research suggests a positive correlation between inflation and growth.

The study was conducted with 30 developing countries around the world, of which 16 are from Asia (the continent with the largest number of developing countries, accounting for 53.33% of the sample), including Cambodia, Indonesia, Laos, Myanmar, Philippines, Thailand, Vietnam, Iran, Iraq, Bangladesh, Egypt, India, Jordan, Nepal, Uzbekistan, and Mongolia. Two European developing countries, Albania and Ukraine, comprise 6.68% of the sample. These are the only developing countries in Europe, as this continent is home to many of the world's most developed economies. Six developing countries in the sample are from the Americas (accounting for 23.33%), namely Colombia, Ecuador, Jamaica, Nicaragua, Paraguay, Peru, and Haiti. Ghana, Namibia, Honduras, Cameroon, and Chad are developing countries from Africa in the sample, accounting for 16.66%. Table 1 outlines the variables in the model.

Table 1 Table describing variables in the research model

Symbol	Variable name	Explanation	Unit	Correlation	Data source
LGDP	Economic growth	Logarithm of per capita GDP			WB
LIQ	Liquidity	Ratio of liquid liabilities to GDP	%	+	IMF
BAN	Banking system development	Number of commercial bank branches per 100,000 people	%	+	WDI
PRC	Private sector credit	Ratio of private sector credit to GDP	%	+	WDI
STO	Stock market development	Stock market capitalization as a percentage of GDP	%	+	WDI
GEX	Government expenditure	Ratio of government expenditure to GDP	%	+/−	WDI
TRA	Trade openness	(exports value + imports value)/GDP	%	+	WDI
INR	Inflation	Annual inflation rate	%	+/−	WDI

4 Result

4.1 *Analysis of Economic Growth and Financial Development in Developing Countries between 2010 and 2023*

The descriptive statistical method is used to provide an overall assessment of the variables in the model from 2010 to 2023. Table 2 shows that the variables' values exhibit relatively large variation. At the same time, the research data is panel data with a strong balance, consisting of 420 observations from 30 countries over 14 years.

Fig. 1 shows that among the 30 developing countries studied, the two countries with the highest average GDP per capita between 2010 and 2023 are Costa Rica

Table 2 Descriptive statistics results

Variable	Obs	Mean	Std.Dev.	Min	Max
LGDP	420	8.269161	1.193879	2.738104	9.716879
STO	420	25.20832	41.63409	0.000858	355.5192
INR	420	8.179041	17.63138	−3.74915	221.3416
GEX	420	130.5594	682.9329	0.739671	4388.202
PRC	420	59.66882	39.29067	1.203972	194.674
BAN	420	13.00695	9.173316	0.348314	48.24884
LIQ	420	75.76797	51.88007	14.01167	268.569
TRA	420	71.91097	34.07993	22.77218	186.6758

Note. Author's analysis

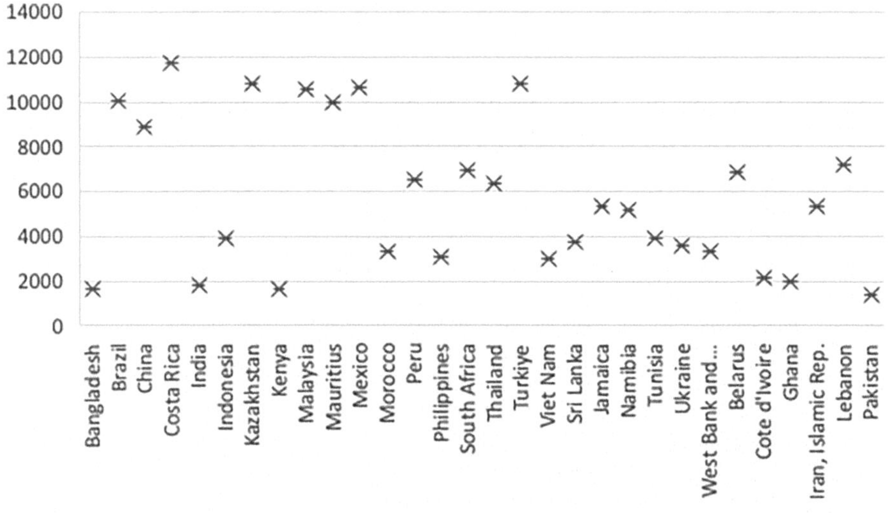

Fig. 1 The average GDP per capita of countries in the study sample for the period 2014–2023. *Note.* Source: Author's analysis based on World Bank data (2010–2023)

($11,712 per person) and Turkiye ($10,851 per person). In contrast, the average income of most other countries in the sample falls between $2000 to $4000 per person. Pakistan has the lowest GDP per capita, at only $1380 per year, and it is a country that has low productivity, political instability, energy shortages, and high levels of corruption and debt. Ranked above Pakistan is Kenya, with a GDP per capita of $1640 per person. Among Southeast Asian developing countries, Vietnam has the lowest GDP per capita at $2981 per person.

From 2010 to 2023, the period saw significant development and fluctuations in financial indicators across developing countries, reflecting the efforts and challenges of building a robust financial system. From the liquidity debt-to-GDP ratio, the number of bank branches per 100,000 people, to the share of private sector credit and market capitalization of the stock market to GDP, each indicator shows uneven development among countries, presenting both opportunities and potential risks. These changes reflect a transformation in economic structures and require countries to adopt effective management strategies to ensure financial stability and promote sustainable economic growth. Specifically:

The Ratio of Liquid Liabilities to GDP During this period, the ratio of liquid liabilities to GDP in 30 developing countries showed significant fluctuations, with most countries trending upwards. Nations like China, Lebanon, Mauritius, Morocco, Vietnam, Thailand, Malaysia, and Brazil had high financial depth (over 100%), indicating a well-developed financial system but posing risks to macroeconomic stability. Conversely, countries like Indonesia, Kazakhstan, Kenya, Mexico, Sri Lanka, and Ivory Coast maintained a low and stable ratio (below 50%), reflecting cautious financial strategies but potentially limiting economic growth due to insufficient financial resources.

Bank Branches per 100,000 People The number of bank branches per 100,000 people fluctuated, with many countries experiencing a decrease due to the rise of digital finance solutions. Sri Lanka, Lebanon, Mauritius, Morocco, Iran, and Tunisia had high bank branch coverage (over 20%), indicating a concentration of large financial institutions, which brings both efficiencies and risks. Countries like Belarus, Kazakhstan, Ukraine, and Vietnam had very low branch coverage (under 5%), highlighting the need for strategies to promote financial inclusion.

The Ratio of Private Sector Credit to GDP Private sector credit as a share of GDP showed fluctuations with a slight upward trend. Countries like China, Malaysia, Lebanon, South Africa, Thailand, and Vietnam had high shares (over 100%), suggesting robust financial systems but potential bad debt risk. In contrast, Ivory Coast, Ghana, Pakistan, and Sri Lanka had low shares (under 30%), indicating restricted credit access and a need for financial reforms.

The Ratio of Stock Market Capitalization to GDP Stock market capitalization to GDP varied significantly, with countries like South Africa, Thailand, Brazil, Malaysia, Sri Lanka, Turkey, and China exhibiting high ratios (over 100%), indicating strong development but vulnerability to financial shocks. Most other countries

Table 3 Correlation matrix

	LGDP	STO	INR	GEX	PRC	BAN	LIQ	TRA
LGDP	1.0000							
STO	−0.2001	1.0000						
INR	0.0174	0.0010	1.0000					
GEX	−0.8173	0.4048	−0.0525	1.0000				
PRC	0.4521	0.4345	0.0394	−0.2707	1.0000			
BAN	−0.3386	0.2311	0.0911	0.5707	0.0073	1.0000		
LIQ	0.3161	0.2906	0.2238	−0.2074	0.7945	0.1548	1.0000	
TRA	0.2907	−0.1491	0.0944	−0.2018	0.3849	−0.2621	0.2599	1.0000

Note. Author's analysis

showed low ratios (below 1%), highlighting underdeveloped markets and the need for transparency and capital mobilization reforms.

In general, the uneven development of financial indicators shows that each country needs flexible and comprehensive strategies to promote the sustainable development of the financial system while effectively managing potential risks (Table 3).

4.2 Empirical Findings

Table 3 presents the correlation matrix of the variables in the model.

Technically, the panel data may have group effects, time effects, or both. These effects can be fixed effects or random effects. The Hausman test is performed to find whether the fixed effects model (FEM) or random effects model (REM) is suitable. The result shows that the FEM is more appropriate than the REM.

Testing for Model Misspecification

Multicollinearity

The author uses variance inflation faction (VIF) to analyze the multicollinearity. If Mean VIF > 10 or 1/VIF < 0.1 indicates trouble. In this case, Mean VIF = 2.74, indicating that the model has no multicollinearity.

Heteroscedasticity: (Wald test)

The Wald test is an effective way to detect heteroskedasticity in a fixed-effect model. The null hypothesis is that residuals are homoskedastic. We reject the null at 95% because Prob > chi2 = 0.000 or a significant Wald test.

Autocorrelation

Another issue with multiple regression models is autocorrelation. To test this issue, the author employs the Arellano–Bond test. The null hypothesis is that there is no first-order autocorrelation. In this case, Prob > F = 0.0000, we reject the null hypothesis, or there is autocorrelation.

The feasible Generalized Least Squares (GLS) is the right choice to deal with the issues of heteroskedasticity and autocorrelation. The GLS regression results are presented in Table 4.

Table 4 Overview of the GLS regression analysis results

Estimated covariances = 30 Estimated autocorrelations = 1 Estimated coefficients = 8				Number of obs = 420 Number of groups = 30 Time periods = 14 Wald chi2(7) = 94.12 Prob > chi2 = 0.000		
LGDP	Coef.	Std. Err.	z	P > z	[95% Conf. Interval]	
STO	−0.00020	0.00030	−0.67000	0.50600	−0.00079	0.00039
INR	−0.00320	0.00078	−4.13000	0.00000	−0.00473	−0.00168
GEX	−0.00115	0.00018	−6.41000	0.00000	−0.00150	−0.00080
PRC	0.00469	0.00117	4.00000	0.00000	0.00239	0.00698
BAN	0.00992	0.00429	2.31000	0.02100	0.00151	0.01833
LIQ	−0.00232	0.00098	−2.37000	0.01800	−0.00423	−0.00040
TRA	0.00162	0.00068	2.38000	0.01700	0.00029	0.00296
_cons	8.07208	0.09127	88.44000	0.00000	7.89319	8.25096

5 Discussions

Results from GLS estimation show that most variables in the model are statistically significant at the 5% level, except for the variable Stock Market Development (STO), which is not statistically significant. Among them, a number of variables are proven to have a positive impact on economic growth, including Private sector credit (PRC), Banking system development (BAN), and Trade openness (TRA). On the other hand, Inflation (INR), Government Expenditure (GEX), and Liquidity (LIQ) are proven to have a negative impact on economic growth.

Liquid Liabilities Considering the GLS estimations methods, the coefficient of LIQ is −0.00232, which means liquid liabilities have a negative effect on economic growth. This is not consistent with the hypothesis. However, it is proved by Gregorio and Guidotti (1995). The result indicates that high monetary liquidity often leads to distortions instead of growth. In many developing economies, increased money supply relative to GDP reflects financial inefficiencies, such as excessive credit creation causing inflation or speculation rather than productive investment. A high money-to-GDP ratio may suggest that savings are poorly directed away from growth sectors. Weak financial institutions, limited access to formal services, and poor credit allocation worsen this issue. External shocks, including volatile markets and the COVID-19 pandemic, further complicate challenges as governments and central banks use expansionary policies that do not guarantee sustainable growth. This highlights the need for improving financial sector efficiency to ensure liquidity fosters productive economic development activities.

Stock Market Development STO's p_value is 0.506, indicating no significant relationship between stock market development and economic growth, contrary to the hypothesis supported by Ayadi et al. (2015). The result suggests these markets might not have integrated into the broader economy. Many developing country stock markets face low liquidity, limited participation, and concentration in a few large firms, hindering their capital mobilization for growth. Speculative trading often prevails, disconnecting them from real economic activities. Structural issues such as weak financial infrastructure, limited regulatory oversight, and low financial literacy further weaken their economic support role. Additionally, external factors like the COVID-19 pandemic and global economic instability may have masked potential positive effects of stock market development.

Banking Sector Development BAN's coefficient is 0.00992, suggesting that banking development positively impacts economic growth. This supports the hypothesis. Improved banking access boosts financial inclusion, enabling savings deposits and credit access. This promotes entrepreneurship in small and medium enterprises, driving job creation and productivity. A broader branch network lowers transaction costs, mobilizes savings, and ensures efficient financial intermediation for productive investments. In rural areas, banking expansion fosters regional development, reduces income inequality, and integrates these regions into the national economy. Additionally, strong banking networks enhance monetary policy transmission, increase resilience to economic shocks, and support trade and innovation through mobile banking.

Private Sector Credit The PRC coefficient is 0.004469, indicating a positive influence of private sector credit on economic growth. This is consistent with the hypothesis. Access to credit helps businesses finance capital, expand operations, and boost productivity, enhancing economic output. For SMEs, crucial in developing economies, credit alleviates financial constraints and aids job creation. Additionally, it fosters household consumption by financing housing, education, and durable goods. Private sector credit has driven industrialization, export growth, and infrastructure development in countries like China, India, and Malaysia. By effectively mobilizing resources, credit deepening encourages structural transformation and diversification, lessening dependence on volatile sectors like agriculture commodities.

Government Expenditure The GLS estimation coefficient of GEX is −0.00115, indicating that government expenditure negatively impacts economic growth, aligning with the hypothesis. Large spending can crowd out private investment by raising borrowing costs or monopolizing resources needed for private sector growth. Developing nations often struggle with corruption and misallocated funds, directing spending toward non-productive or politically motivated projects instead of infrastructure, education, or healthcare. High government expenditure can lead to unsustainable fiscal deficits, causing inflation or debt crises, as seen in Pakistan, Lebanon, and Ghana. Moreover, excessive dependence on public sector growth can hinder competitiveness in the private sector innovation.

Trade Openness The GLS estimations show the TRA coefficient as 0.00162, indicating a positive effect of trade openness on economic growth. This aligns with the hypothesis that open trade policies expand market access, enhance economies of scale, and improve resource allocation efficiency. Exports drive growth by generating foreign exchange, boosting domestic production, and fostering innovation. Imports offer advanced technologies, high-quality inputs, and consumer goods, improving productivity and living standards. Countries like China, Vietnam, and Malaysia have used trade openness to attract foreign direct investment (FDI) and move toward higher-value industries. Additionally, trade openness has helped countries such as Kenya, India, and Morocco diversify their economies, reducing reliance on volatile sectors like agriculture and commodities growth.

Inflation Rate INR's coefficient is −0.0032, indicating that inflation negatively affects economic growth, aligning with the hypothesis. High inflation decreases real income, especially for low- and middle-income families, who allocate more to essentials. For businesses, it raises production costs, disrupts financial planning, and reduces competitiveness domestically and internationally. Ongoing inflation discourages savings and long-term investments due to declining money value. Furthermore, it weakens monetary policy, forcing central banks to implement contractionary measures, like raising interest rates, which can hinder economic activity and growth. Thus, controlling inflation is vital for sustainable economic development in developing countries' economies.

6 Conclusion

This study aims to investigate the effect of financial development on economic growth in 30 developing countries worldwide. Specifically, it examines the effect of financial system development (including stock market development, liquid liabilities, banking sector development, and private sector credit) on economic growth from 2010 to 2023 using GLS estimation techniques.

This study expands the scope of previous studies by focusing on developing economies in Africa (12 countries) while also studying other regions such as Asia, Europe, the Americas, and MENA. Unlike Ho (2018), who focused only on Southeast Asia, or Estrada et al. (2010), who analyzed 116 developing countries mainly in Asia, this study emphasizes the African context in the impact of financial development on economic growth. In addition, the study also extends the global approach of Narayan and Narayan (2013) but the countries come from more different continents.

The research results indicate that financial development positively impacts economic growth in developing countries during the 2010–2023 period, as evidenced by indicators such as the development of the banking system and private sector credit. In these developing countries, a higher number of bank branches and growth in credit lead to increased access to capital and investment for the private sector, reflecting the development of the financial system and contributing to economic growth.

On the other hand, liquidity debt fluctuates in the opposite direction of economic growth in developing countries. To reduce dependence on sectors that are more vulnerable to financial shocks, these countries need to adopt cautious monetary policies, improve debt management, strengthen financial capacity, and diversify the economy by developing new industries and services.

The development of the stock market is statistically insignificant, even though the stock market is an important funding and investment channel. This indicates that in developing countries, there is a pressing need for more practical policies to develop the market in depth, helping it to integrate more closely with the overall economy. Countries must raise transparency standards and strengthen market regulation to ensure a fair and sustainable investment environment. They should also enhance international cooperation to exchange and learn from best practices, attract foreign investors, and adopt technological solutions such as blockchain and artificial intelligence to improve market efficiency and security.

References

Arcand, J. L., Berkes, E., & Panizza, U. (2015). Too much finance? *Journal of Economic Growth, 20*, 105–148.

Ayadi, R., Arbak, E., Naceur, S. B., & De Groen, W. P. (2015). Financial development, Bank efficiency, and economic growth across the Mediterranean. In R. Ayadi, M. Dabrowski, & L. De Wulf (Eds.), *Economic and social development of the southern and eastern Mediterranean countries* (pp. 219–233). Springer International Publishing.

Bayar, Y. (2014). Financial development and economic growth in emerging Asian countries. *ASS, 10*, 8.

Belloumi, M. (2014). The relationship between trade, FDI and economic growth in Tunisia: An application of the autoregressive distributed lag model. *Economic Systems, 38*, 269–287.

Bist, J. P. (2018). *Financial development and economic growth: Evidence from a panel of 16 African and non-African low-income countries*. Cogent Economics & Finance.

Caporale, G. M., Rault, C., Sova, A. D., & Sova, R. (2015). Financial development and economic growth: Evidence from 10 new European Union members. *International Journal of Finance & Economics, 20*, 48–60.

Christopoulos, D. K., & Tsionas, E. G. (2004). Financial development and economic growth: Evidence from panel unit root and cointegration tests. *Journal of Development Economics, 73*, 55–74.

Ciftci, D., Ispir, M. S., & Yetkiner, H. (2017). Financial development and economic growth: Some theory and more evidence. *Journal of Policy Modeling, 39*, 290–306.

Čihák, M., Demirgüç-Kunt, A., Feyen, E., & Levine, R. (2012). Benchmarking financial systems around the world. *The World Bank Policy Research Working Paper*.

De Gregorio and Guidotti (1995). "Financial development and economic growth", *World Development*, Vol 23 No.3, pp. 433–448.

Demetriades, P. O., & Rousseau, P. L. (2016). The changing face of financial development. *Economics Letters, 141*, 87–90.

Estrada, G. B., Park, D., & Ramayandi, A. (2010). *Financial development and economic growth in developing Asia*. SSRN Papers.

Grassa, R., & Gazdar, K. (2014). Financial development and economic growth in GCC countries: A comparative study between Islamic and conventional finance. *International Journal of Social Economics, 41*, 493–514.

Hassan, M. K., Sanchez, B., & Yu, J.-S. (2011). Financial development and economic growth: New evidence from panel data. *The Quarterly Review of Economics and Finance, 51*, 88–104.

Ho, T.Y.N. (2018). "Financial Development and Economic Growth in Some Developing Southeast Asian Countries," *Master's Thesis, University of Economics Ho Chi Minh City.*

Huang, H.-C., & Lin, S.-C. (2009). Non-linear finance–growth nexus. *The Economics of Transition, 17*, 439–466.

Hui, A., Qianmiao, Z. & Mohamed, K. (2021). "Impact of financial development on economic growth: Evidence from Sub- Saharan Africa", *Australian Economic Papers*, Vol. 60 No. 2, pp. 226–260.

Hye, M. (2012). Exports, imports and economic growth in China: An ARDL analysis. *Journal of Chinese Economic and Foreign Trade Studies, 5*, 42–55.

Ibrahim, M., & Alagidede, P. (2018). Effect of financial development on economic growth in sub-Saharan Africa. *Journal of Policy Modeling, 40*, 1104–1125.

Jedidia, B. K., Boujelbène, T. D., & Kamel, H. (2014). Financial development and economic growth: New evidence from Tunisia. *Journal of Policy Modeling, 36*(5), 883–898.

Jude, C. (2010). Financial development and growth: A panel smooth regression approach. *Journal of Economic Development, 35*, 15–33.

Khan, M. S., & Senhadji, A. S. (2003). Financial development and economic growth: A review and new evidence. *Journal of African Economies, 12*, ii89–ii110.

Kimaro, E. L., Keong, C. C., & Sea, L. L. (2017). Government expenditure, efficiency and economic growth: A panel analysis of Sub Saharan African low income countries. *African Journal of Economic Review, 5*, 34–54.

Le, P.T.Q & Nguyen, T.P.T. (2021). "The Impact of Financial Development on Economic Growth in ASEAN Countries", *Journal of International Management and Economics*, Vol 145, tr. 53–68.

Mishra, P.K., Das, K.B. & Pradhan, B.B. (2009). "Credit market development and economic growth in India", *Middle Eastern Finance and Economics*, Vol. 5 No. 3, pp. 92–106.

Narayan, P. K., & Narayan, S. (2013). The short-run relationship between the financial system and economic growth: New evidence from regional panels. *International Review of Financial Analysis, 29*, 70–78.

Pradhan, R. P., Arvin, M. B., Bahmani, S., Hall, J. H., & Norman, N. R. (2017). Finance and growth: Evidence from the ARF countries. *The Quarterly Review of Economics and Finance, 66*, 136–148.

Rajan, R. G. (2005). Has financial development made the world riskier. NBER working paper series, Working Paper 11728.

Romer, P. M. (1986). Increasing returns and long-run growth. *Journal of Political Economy, 94*, 1002–1037.

Rousseau, P. L., & Wachtel, P. (2002). Inflation thresholds and the finance–growth nexus. *Journal of International Money and Finance, 21*, 777–793.

Salahuddin, M., & Gow, J. (2016). The effects of internet usage, financial development and trade openness on economic growth in South Africa: A time series analysis. *Telematics and Informatics, 33*, 1141–1154.

Trinh, T.N.G. (2018). "The Relationship between Financial Development and Economic Growth in Some Asia-Pacific Countries," *Master's Thesis, University of Economics Ho Chi Minh City.*

Thornton, J., & Poudyal, S. R. (1990). Money and Capital in Economic Development: A test of the McKinnon hypothesis for Nepal: Note. *Journal of Money, Credit and Banking, 22*, 395–399.

Yang, F. (2019). The impact of financial development on economic growth in middle-income countries. *Journal of International Financial Markets, Institutions and Money, 59*, 74–89.

Zhang, J., Wang, L., & Wang, S. (2012). Financial development and economic growth: Recent evidence from China. *Journal of Comparative Economics, 40*, 393–412.

The manufacturer's authorised representative in the EU is Springer Nature Customer Service Centre GmbH, Europaplatz 3, 69115 Heidelberg, Germany. If you have any concerns regarding our products, please contact ProductSafety@springernature.com

Printed and bound by CPI Group (UK) Ltd, Croydon, CR0 4YY
01/10/2026
02225925-0006